For my Father.

"When we approach the **great philosophical systems** of Plato or Aristotle, Descartes or Spinoza, Kant or Hegel, with the criteria of precision set up by mathematical logic, these systems fall to pieces as if they were houses of cards. Their basic concepts are not clear, their most important theses are incomprehensible, their reasoning and proofs are inexact, and the logical theories which often underlie them are practically all erroneous. Philosophy must be reconstructed from its very foundations; it should take its inspiration from scientific method and be based on the new logic. No single individual can dream of accomplishing this task. This is work for a generation and for intellects much more powerful than those yet born"
- Jan Lukasiewicz.

"And now empiricist scepticism brings to light what was already present in the Cartesian fundamental investigation but was not worked out, namely, that all knowledge of the world, the pre-scientific as well as the scientific, is an enormous enigma."
- Edmund Husserl.

1. The Radiation Origin. 5
2. The Quantum Origin. 7
3. The Transition Problem. 8
4. The Riddle of the universe. 10
5. The Zero Sum Universe. 11
6. Universal Relativity. 12
7. Is Entropy Conserved? 14
8. Is Time Curved? 16
9. The Physical Significance of Curved Time. 17
10. Hyper-symmetry. 19
11. The Engineering Functions of the Universe. 21
12. Pierre Curie and the Principle of Sufficient Reason. 22
13. Open Or Closed? 25
14. The Critique of Pure Inflation. 27
15. Net Perfect Order. 31
16. The Physical Significance of Net Perfect Order. 32
17. An Imaginary Cosmology. 34
18. The Hyper-Atom. 35
19. The Rebirth of Rationalism? 37
20. Back to the Eleatics. 38
21. The Rational Basis of Inductive Philosophy. 41
22. Zeno and Quantum Physics. 42
23. Why Zeno's Paradoxes are Both Veridical and Falsidical. 44
24. The Super-Universe. 45
25. Is the Universe in a Spin? 46
26. Black-holes Unmasked. 48
27. Baryothanatos. 50
28. The Conservation of Information. 51
29. The Third Law of Information Theory. 53
30. Indeterministic Order. 54
31. The Correspondence Principle. 56
32. Concerning Incommensurability. 57
33. The Problem of Causation. 59
34. The Physical Significance of Indeterminate Order. 61
35. Deterministic Chaos. 64
36. Why the Anthropic Principle is not even Wrong. 66
37. The Measurement Problem. 67
38. The Nihilistic Interpretation of Quantum Mechanics. 68
39. Wonderful, Wonderful Copenhagen. 70
40. Of Single Photons and Double Slits. 71
41. My Brane Hurts. 72
42. M-Theory and Empiricity. 73
43. The Problem of Decidability. 75
44. The Logic of Nature? 77
45. The Foundations of Logic. 79
46. The Quine-White Uncertainty Principle. 80
47. The Rebirth of Analytical Philosophy. 82
48. The Incompleteness of Empiricism. 85
49. The Epistemological Basis. 87

50. Three-Valued Logic. 89
51. Pyrrho and the End of Greek Philosophy. 91
52. The Second Truth Theorem. 95
53. Neo-Logicism. 97
54. What is Truth? 101
55. Naturalized Epistemology and the Problem of Induction. 103
56. The Problem of Epistemology. 105
57. The Correct Interpretation of Instrumentalism. 107
58. The Roots of Phenomenology. 108
59. Ontological Commitment versus Hyper-Nihilism. 110
60. Non-classical Completeness. 112
61. Gödel's Uncertainty Principle. 114
62. The basis of Rational Philosophy. 115
63. Two types of Indeterminacy. 116
64. The Trivalent Basis of Induction. 118
65. Neo-Foundationalism. 119
66. Induction and Revolution. 120
67. The Rendering of Classical Philosophy. 122
68. The Trivalent Foundations of Mathematics. 123
69. The Foundations of the Apriori. 124
70. The Limits of Phenomenology. 126
71. What is Metaphysics? 128
72. The Problem of Ontology. 129
73. The Ontological Deduction. 130
74. Neo-Rationalism. 133
75. The Essence of Metaphysics. 134
76. The Roots of Philosophy. 136
77. The Illusion of Syntheticity. 138
78. Plato's Project. 139
79. The Triumph of Rationalism. 141
80. The Modern synthesis. 142
Appendix; Prime Numbers, Indeterminism and the Riemann Hypothesis._____145

1. The Radiation Origin.

The universe tunnels into existence at 10^{-43} seconds after the beginning. High energy photons, which are the source of all matter, rapidly condense into plasma some-time after 10^{-36} seconds. At the conclusion of a series of phase transitions involving the disaggregation of the fundamental forces – gravity, the two nuclear forces and electromagnetism – protons and neutrons form out of the dense quark sea at around 10^{-6} seconds A.T.B. Over the next few minutes, hydrogen and helium nuclei synthesize as the overall temperature further declines. However, it is not until around another 300,000 years have passed that temperatures become cool enough (at around $3000°^k$) to allow these nuclei to collect electrons and so form electrically neutral atoms. At this point, for the first time, matter becomes visible due to the reduction in the sea of free electrons. Approximately a billion years later (it is suspected) quasars, formed out of these atomic nuclei, begin decaying into galaxies, leaving a universe much as it is at present, albeit considerably denser and smaller in extent.

This, in its most general form, is the current cosmological model for the origin and evolution of our universe (one which is still taking distinct shape of course), a remarkable model which is a product of the empirical method, of induction and deduction based on observation and experience.

To establish induction as a legitimate source of ontology and not merely as a method (or even as a form of epistemology) one problem in particular has to be dealt with satisfactorily in empirical terms and that is the problem of the origin of the universe itself. To help us with this problem I believe that it is crucial to distinguish between two different yet related conceptions of the origin.

The first and most obvious conception is the "classical" or "radiation" origin of the universe where time equals zero (T=0) and nothing may be said to exist at all (in fact it is characterized by a condition of perfect indeterminacy). Though currently a classical concept defined in terms of the *General Theory of Relativity* and commonly known as the "singularity", I believe that the radiation origin is susceptible to a simple yet illuminating quantum description that does away with the abnormalities inherent in the relativistic or "classical" account.

Applying the analysis of General Relativity at the radiation origin is, in any case, wrong and not simply because it results in meaningless predictions of infinities for curvature and matter density. General Relativity indicates to us the *existence* of the radiation origin, but it cannot tell us anything beyond this. In other words it has *no* ontological significance. The correct description of the radiation origin can *only* be defined by means of quantum theory since this is the only appropriate theory for the sub-atomic scale. By insisting on applying an inappropriate, i.e. "classical", theory at this scale it is unsurprising that we should therefore end up with a meaningless result.

So what does quantum theory (which has never before been applied here) tell us about the radiation origin? I believe that quantum theory provides us with a description which is (as should be expected) both simple and clear. Under this analysis the radiation origin is defined as a "particle" with zero wavelength and hence *indeterminate* potential energy. This, incidentally, describes a state of *perfect order* or *symmetry* since there is also no information and hence no entropy at this point.

The correct interpretation of this discovery is that the radiation origin is a ***perfect black-body***, which is to say; a state of affairs capable of radiating energy across a spectrum of wavelengths from zero to infinity. This further corroborates the view that the correct treatment of the radiation origin *must* be a quantum theoretical one since quantum theory arose historically as a means of accounting for the (at that time) anomalous properties of perfect black-body radiators such as this. It therefore stands to reason that the solution to the conundrum of the radiation origin – a problem raised by classical physics but not *resolvable* by it – can only be a quantum mechanical one. The relevant quantum equation to describe the radiation origin in terms of quantum theory (which has not been cited in this context before) is, I believe, the following;

$$E = \frac{h\nu}{\lambda}$$

At the radiation origin λ (wavelength) and ν (velocity) are *both* zero, therefore leading to *indeterminacy*, (since a zero divided by *another* zero gives an *indeterminate* rather than an infinite result) in effect contradicting (and supplanting) the classical or relativistic account. The energy of the universe at T=0 may therefore radiate from zero to infinity according to the quantum analysis, thereby proving that the radiation origin *must* be a perfect black-body radiator. The radiation origin must in any case be indeterminate (and thus a perfect black-body) since any other condition would convey information and hence imply entropy. This in turn would beg the question as to the origin of the entropy.

It could fairly be argued that the values for wavelength and velocity are *also* indeterminate at this point rather than being zero. Though this is true it merely serves to validate our general argument that *all* values at the radiation origin are indeterminate rather than infinite and that the correct analysis of this point must be a quantum mechanical rather than a relativistic one. This point is, I believe, inescapable and profoundly significant.

In any event, it does not really matter if the classical and quantum descriptions of the radiation origin fail to agree since the quantum description provided by this equation has indisputable epistemological primacy in this case, (thus resolving any contradictions). This is because cosmology in the very early universe *is* particle physics and nothing else. And what it describes (ironically given the historical genesis of quantum theory) is quite clearly *not* a "singularity" (as according to contemporary orthodoxy) but a perfect black-body.

One might reasonably ask what difference this makes since both descriptions, the classical and the quantum, seem to be quite similar. But this is just a superficial appearance. In reality, as Hawking and Penrose have demonstrated, *General Relativity* predicts infinities, which are impossible, whereas quantum theory predicts *indeterminacy*, which is the stuff of regular day to day quantum mechanics. Indeed, as just mentioned, it was precisely the problem of the perfect black-body radiator (ipso-facto a state of perfect order) that led Max Planck to first postulate the quantum theory in 1900. It seems that, unlike singularities, perfect black-bodies *do* submit to some sort of quantum resolution and the radiation origin (as we shall see in the next two sections) is no exception. It is also a result that proves once and for all what has long been suspected; that relativistic singularities (whether of the Hawking-Penrose or the Schwarzschild varieties) are *never* possible, in this or any other circumstance[1].

The relativistic prediction of infinities and infinities alone for this point is in effect rubbed out by this analysis, which is a significant advance in its own right since it restores analytical rationality to this most important time. *Furthermore, the recent discovery that the cosmic back-ground radiation displays a perfect black-body spectrum represents effective experimental confirmation of these conclusions*, which are (as evidenced by the above equation) that the radiation origin *is* a perfect black-body and *not* a singularity.

---

[1] If I would like my work on physics to achieve one thing it would be the elimination of the erroneous and unnecessary "singularity" concept from physics. In its place should go a full understanding of the concept – at once ontological and epistemological – of *indeterminacy*. It is this "full understanding" (including an account of the clear and fundamental role played by indeterminacy in generating order and regularity as limiting cases) which constitutes the ambition and common thread of this work.

The uncertainty principle does, in any case, forbid the precision with which the singularity is defined.

2. The Quantum Origin.

In general then the radiation origin decomposes into two separate and contradictory descriptions, a classical and a quantum one, of which the latter has indisputable priority. The second conception of origin I wish to introduce, the quantum origin, admits, as its name suggests, of only a *quantum* description.

The precision of the description of the universe at what I call the quantum origin (distinguished by the fact that it is the earliest point of *determinate* time) is quite remarkable albeit entirely the product of quantum theory. Notwithstanding this theoretical character, what is described at this point is, perhaps surprisingly, a subatomic particle, albeit one without precedent in particle physics. As such it would not be unreasonable to characterize this "super-particle" as a new state of matter. In this section I will discuss some of its defining characteristics.

The three fundamental constants, for gravity, the velocity of light and Planck's constant, allow us to define nugatory values for time (as we have seen) but also for length (approximately $10^{-33} c.m.$) and for mass (approximately $10^{-5} kilos$). These "Planck dimensions" are the most miniscule allowable by quantum theory and so are normally interpreted as representing fundamental structural limits in nature itself. These three constants also give rise to a fundamental unit of energy known as the Planck energy whose value is approximately $10^{19} Gev$ or $10^{32°k}$. This represents, I believe, the "ceiling temperature" of the universe attained only at the quantum origin itself.

The fundamental dimensions suggest to me a "particle" at the quantum origin ($10^{-43} \sec onds$) with a diameter of $10^{-33} c.m.$ and a density of $10^{93} g/cm^3$ (the Planck density). These dimensions seem to be a rational and even an inevitable hypothesis and clearly define a new species of particle which is ipso-facto a new state of matter. Incidentally, the justification for so called "string theory" is somewhat similar since it too founds itself on the fundamental character of the Planck dimensions and is therefore very certainly a quantum theory through and through.

The quantum origin is a new species of particle because it cannot be fitted into any already existing category. Consider the facts. Since this particle is mass bearing (with a mass of $10^{-5} kilos$) it *must* be characterized as being, by definition, a fermion with spin ½ - since all mass-bearing particles have these characteristics. The only other known fermions are electrons and quarks, but this "super-particle" cannot be ascribed to either of these categories. Unlike the electron (or its anti-type the positron) we have no grounds to ascribe any electrical charge to it. It cannot be characterized as a quark either since quarks always occur in pairs or triplets with part charges. Therefore it follows that this briefly existing state of affairs at the quantum origin (which decays in one Planck time) counts as a new state of matter and a new particle type.

The sheer precision of these statistics at the quantum origin raises a problem however. Their precision appears to transgress the time/energy uncertainty relation, which is perhaps evidence of a flaw in the theory. The time and the energy for the universe at this point are stated with equal precision. Presumably this astonishing transgression (if the interpretation is correct) is due to the fact that these vital-statistics for the universe at this point are entirely derivable from the quantum theory alone and so do not rely on any invasive measurements of the sort that are interpreted as being ultimately responsible for the uncertainty in quantum mechanics. This, if I am right, is a unique exception however and does not occur again, if, indeed, it occurs at all.

Finally it may be pertinent to remind ourselves of the discovery in 1996 of a strikingly symmetrical circumstance – a new state of matter (called the Bose-Einstein condensate) existing just a few milli-Kelvins above the *basement* temperature of absolute zero. Though not proof of anything this possible symmetry between the two temperature extremes of the universe is still intriguing.

3. The Transition Problem.

This account of the two origins presents us with the question; by what means does the universe transition from the radiation origin to the quantum origin? The solution to this seemingly intractable problem is simple and categorical and supplied by quantum mechanics in the form of the uncertainty principle. This fact is further evidence that quantum theory is indeed the correct formalism with which to describe the two conceptions of origin.

Where time is precisely known, as it is at the radiation origin, the uncertainty principle indicates that energy becomes equivalently uncertain and can therefore amount to anything from zero to infinity – very much in line with our understanding of the radiation origin as a perfect black-body radiator. Theoretically and for reasons we shall discuss in the next section, energy *must* take on a finite, non-zero value and, in practice, this takes the form of the quantum origin previously discussed.

Conversely, where energy is precisely stated our knowledge of time becomes correspondingly destabilized. And so the universe can be engendered, ex-nihilo, according to the well defined rules of quantum mechanics. If there are any doubts about this they can perhaps be dispelled with reference to the following uncertainty relation;

$$\Delta T \cdot \Delta E \geq \hbar$$

Which shows us that when T=0 (the radiation origin of the universe) the uncertainty in the energy state of the universe at this point (ΔE) becomes infinite;

$$\Delta E = \frac{\hbar}{0} = \infty$$

Thus allowing for the transition to the quantum origin.

The solution to what I call the "transition problem" is not new however. It was first mooted, in a slightly different form, in 1973 by the American physicist Ed Tryon in an article in "Nature" magazine entitled "Is the Universe a Vacuum Fluctuation"[2]. Now obviously Tryon's discovery of this solution has priority over my own which was made independently some twenty years later, however Tryon's version raises a couple of serious problems that are not present in the above restatement. First and most importantly, Tryon, to put it bluntly, adduces the wrong version of the uncertainty principle in support of his hypothesis. He uses the familiar position/momentum uncertainty relation whereas the less well known time/energy version is clearly more apposite to the case being dealt with (that of the origin of the universe). Neither position nor momentum (nor energy for that matter) can be established for the radiation origin (making Tryon's citation of questionable judgment). However, *time* at the radiation origin can by definition be said to equal zero. The correct application of the uncertainty principle can therefore only be the time/energy one presented above.

Secondly, Tryon couches his insight in the conventional context of the "quantum vacuum". This concept which was originated in the 1930s by Paul Dirac describes the *normal* context in which energy can be spontaneously created out of nothing in empty space. In the extraordinary context of the radiation origin however its use becomes ontologically suspect. This is because at this point there is no space or time and so there can be no vacuum either. The vacuum is a product of these events and not vice-versa. Since the

---

[2] Tryon, Edward P. "Is the Universe a Vacuum Fluctuation," in *Nature*, 246(1973), pp. 396-397.

radiation origin is void of time and space it therefore seems more accurate to speak of this unique situation, not as a quantum vacuum (as Tryon rather unquestioningly does) but rather (in order to make this important distinction) as a *quantum void*. The mechanics of the fluctuation remain identical but the terminology used to describe it should surely be altered, along the lines suggested here.

4. The Riddle of the universe.

The transition problem is solved by the uncertainty principle, which is the means by which the perfect symmetry of the radiation origin is decisively broken. But the *motivation* for this transition is still unexplained. Why, in other words, is the perfect order of the radiation origin sacrificed for the disequilibrium of the quantum origin? This is not only the most fundamental question of cosmology; it is, implicitly, also the most fundamental question of ontology as well. To answer it is to answer Martin Heidegger's well known question; "Why does anything exist at all? Why not, far rather, nothing?"[3]

To sustain perfect order the universe must manifest either zero or else infinite energy. Any other solution results in disequilibrium and hence disorder. Being mathematical abstractions however neither of these solutions possesses physical or thermodynamical significance. Consequently, the universe is obliged to manifest finite, non-zero properties of the sort associated with the quantum origin. The radiation origin, in other words is *mathematically* perfect, but physically meaningless. The quantum origin, arrived at quantum mechanically, is the inevitable outcome of this limitation.

Nature, it seems, steers a course so as to avoid the Scylla of zero and the Charybdis of infinite energy. And this fact accounts for both the *existence* of the universe and also for its discrete or quantum format. The specifically *quantum* format of the universe, to be more precise, is explicable as the most *efficient* means by which the "ultra-violet catastrophe" posed by the radiation origin can be avoided. It is *this* fact which answers John Wheeler's pointed and hitherto unanswered question "Why the quantum?"

It is therefore reasonable to conclude that the universe is the outcome of logical necessity rather than design or random chance. Nature is conveniently "rational" not for transcendental or apriori reasons but because this is the least wasteful of solutions. Creation ex-nihilo is the most efficient solution to the paradox of the radiation origin, ergo nature opts for it. Indeed, to digress only slightly, logic itself is valid *because* it is efficient, even though it, like everything else, lacks absolute foundation (as has been implied by the discoveries of Kurt Gödel and Alonzo Church).

The universe therefore exists as a *thermodynamical* solution to a *logical* problem – to wit that *something* (in practice mass-energy) *must* exist in lieu of the self-contradicting abstraction "absolute nothingness". Consider, after all, the paradox contained in the idea of the *existence* of absolute nothingness, i.e. the existence of inexistence. Nothingness, by definition, can never be *absolute;* it has, instead, to take *relative* form – and this, of course, is what our universe does.[4]

This view of course implies the apriori character of the space-time continuum, a view in accord with Kant's concept of the "forms of intuition" (i.e. space and time) which he interprets as possessing a distinctive apriori character.

---

[3] Martin Heidegger "What is Metaphysics?" 1926. See also; Gottfried Von Leibniz. "On the Ultimate Origination of Things." 1697.
[4] Another way of putting this is to observe that though the root of being may be nothingness nevertheless the *consequence* of nothingness must logically be being. Thus it is possible to argue that existence is an apriori concept *independent of experience,* a view which accords with Kant and the Rationalists but which seems to refute the basic premise of empiricism; no concept without experience.

5. The Zero Sum Universe.

I have hopefully demonstrated that the correct description of the radiation origin is not that of the "singularity" supplied by general relativity, but rather, of a perfect black-body as described by quantum theory (reference the equation in section one). This result, furthermore, explains the *format* of the universe, which is quantized so as to solve the black-body problem (i.e. the problem of the ultra-violet catastrophe) posed by the radiation origin. Hitherto the *quantum* form of the universe was inexplicable, but a correct quantum description of the radiation origin (i.e. as a perfect black-body) decisively clarifies this ontological mystery. A further consequence of this discovery is that the universe can be characterized as the outcome of black-body (in effect *vacuum*) radiation.

The universe therefore expands on vacuum energy since the vacuum or *quantum void* has unlimited potential energy. All energy is thus, at bottom, vacuum energy, which is to say; energy conjured out of nothingness (since nature does indeed *abhor a vacuum*) according to the mechanics of the quantum fluctuation discussed earlier.

As particle cosmologist Professor Alan Guth has pointed out a quantum fluctuation at T=0 is a far more consequential phenomenon than an equivalent such event today – which is invariably brief and insignificant. A quantum fluctuation at the radiation origin, by contrast, is of potentially infinite duration;

> "In any closed universe the negative gravitational energy cancels the energy of matter exactly. The total energy or equivalently the total mass is precisely equal to zero. With zero mass the lifetime of a quantum fluctuation can be infinite."[5]

This observation, implying the conservation of *zero* energy over the duration of the universe, raises issues concerning our interpretation of the first law of thermodynamics. Since energy cannot be created or destroyed how can energy truly be said to exist at all? Even though the universe *is* replete with many forms of energy they *all* cancel one another out in sum. The obvious and best example of this is positive and negative electrical energy – which cancel each other out precisely. But this principle must also apply, as Guth observes, to mass and gravitational energy as well. Guth's observation therefore supplies the clue to clearing up any mystery surrounding the interpretation of the first law; energy cannot be created or destroyed, not because it does not exist, but due to its ephemeral or *void* quality.

---

[5] "The Inflationary Universe" p272. Alan Guth. Vintage Press, 1997.

6. Universal Relativity.

All three laws of thermodynamics have major implications for ontology, as we shall have further occasion to observe. But what of the other key theory of classical physics, Einstein's relativity?

Relativity per-se can be used as evidence of ontological nihilism. If all phenomena have a relative existence and are defined entirely by their relation to *other* phenomena (an argument reminiscent of Kant's famous "transcendental deduction"), then no entity can be considered substantial. This is a major argument utilized by the Mahayana nihilists in order to demonstrate that nothing has inherent or permanent existence. Known sometimes as the "doctrine of universal relativity" this remarkable "deduction" of ontological nihilism is normally attributed to the Buddha himself.

Conversely, since inexistence cannot be "absolute" (for the logical reasons debated in section four) it must take a *relativistic* form instead. Thus absolute nothingness *necessitates* (takes the form of) relative being.

The problem with the doctrine of universal relativity in this context is that Einstein's relativity is not relative in the same sense, but in a much narrower one. In this instance all *motion* is held to be relative to the velocity of light in a vacuum. Since this velocity is a constant it has been justifiably argued that Einstein's work represents a theory of the absolute, in disguise.

Indeed, this interpretation, which is the correct one, appears to put paid to the more general *Mahayana* interpretation of relativity. However, quantum mechanics explodes this foundation of Einsteinian relativity by revealing Einstein's "absolute" (i.e. the velocity of light in a vacuum) to be an ideality since the "vacuum" referred to never possesses real physical existence. The velocity of light in practice is therefore simply another experimental variable subject to the vagaries of quantum uncertainty and not a real absolute.

Faster than light speed interactions – which are not ruled out by quantum mechanics – would be indistinguishable from "normal" vacuum fluctuations which are observed routinely and predicted by the Dirac equation. Therefore the transgression of Einstein's classical barrier engenders no serious challenge to our current understanding and does not, for example, require the hypothesis of faster than light "tachyons".

Quantum mechanics then, in demonstrating to us that there *are* no absolutes is thus the *true* physical theory of universal relativity (thus quantum mechanics is the most general mechanical theory of all). This is because, as the uncertainty principle demonstrates, it is impossible to measure a phenomenon without simultaneously interfering with it. This represents a clear demonstration of universal relativity at the most fundamental level. In other words, the uncertainty principle is nothing less than an objective formalization of the principle of *universal* relativity. And this is the source of the profound ontological significance of quantum mechanics, (see section thirty eight for a fuller discussion of this key issue).

There is no doubt, furthermore, that universal relativity (which is implied by quantum mechanics, but *not* by Einsteinian mechanics) inescapably entails ontological nihilism (as according to the purely logical argument of Mahayana nihilism) and that onto-nihilism therefore represents the profoundest possible interpretation of quantum mechanics.

Notwithstanding this critique of foundations implied by quantum mechanics, Einstein's relativity does possess a number of implications for ontology. We have already touched upon one such implication in our treatment of the radiation origin. The achievements of Hawking, Penrose and Ellis in cosmology lie, as we have seen, in examining the limits of the general theory of relativity and thereby leading us to our initial conception of the radiation origin. At this point however a quantum description *must* take over since cosmology in the very early universe *is* particle physics, thus explaining what is freely admitted by relativistic cosmologists; that general relativity ceases to be a useful tool at this point.

The key to the nihilistic interpretation of the theory of relativity however is to see that *all* mass is frozen acceleration or frozen energy according to the famous equation $E = MC^2$. This indeed is the significance of Einstein's enigmatic remarks to the effect that all objects are traveling at light velocity. It is simply that this motion (in the form of matter) is divided between time and space. If *all* mass were to be converted into energy (as is the case only at the radiation origin) then there would be *no* frames of reference and so no time or space. There would simply *be* nothing, thus demonstrating onto-nihilism as a hidden corollary of the theory of relativity.

7. Is Entropy Conserved?

The fact that net energy in a closed universe is zero suggests that net entropy must be zero too. But this is not the standard interpretation of the second law of thermodynamics, which is normally formulated as an inequality;

$$\partial S = \frac{\partial q}{T} \geq 0$$

That is; changes in entropy (∂S), which are equivalent to changes in heat flow (∂q) divided by temperature (T), are always greater than (or equal to) zero. This inequality holds under the assumption of a flat or open universe, both of which continue expanding forever (at a diminishing rate in the former case and at an accelerating rate in the latter). In such universes, the non-conservation of entropy does indeed always obtain.

However, in the contracting phase of a closed universe the above inequality must be reversed, indicating a *decrease* in entropy over time, which perfectly balances the increase in the prior expansionary phase, (a closed universe is the sum of these two phases). In a contracting universe it is *negative* entropy – the tendency of temperature to increase its activity through time (implying temperature increase) which should be ascendant.[6]

Entropy reversal is not dependent upon the reversal of time, as was once hypothesized by Professor Stephen Hawking; entropy reversal merely implies greater uniformity (symmetry) and higher temperatures over time, which are the unavoidable consequences of a shrinking universe. Since entropy is no more than the measure of the expansion of heat flow it follows inescapably that as space contracts in a shrinking universe, heat flow must also contract – generating higher temperatures, more work and greater order.

Furthermore the mathematical work of Professor Roger Penrose, upon which Hawking's own relativistic cosmology is based, appears to suggest that the contraction of matter towards a singularity condition (as at the end of a closed universe) generates increasing order and symmetry over time and hence decreasing entropy or heat flow. A singularity, in other words, is an example of perfect symmetry, not complete disorder.

Further confirmation of this supposition is given by the consideration of a collapsed universe entering its final phase, as a black-hole. At this point, according to the work of Hawking and Bekenstein, the universe will possess the *maximum* amount of entropy possible for an object of its size, since this is true of *all* black-holes of *any* size. Nevertheless the universe (incorporating its event horizon) will *continue*

---

[6] The speculations here and in the next few sections give rise, I believe to two powerful deductions that are of great importance in justifying the new cosmological model that I am going to describe;

1) The Closed Universe Deduction; if the first law of thermodynamics is correct then the universe must be closed. This is because this law (that energy can neither be created nor destroyed) effectively requires that there must be net zero energy in the system and this can only be the case in a *closed* universe.

Indeed, Pierre Curie's principle of symmetry implies energy must be net zero since energy is a phenomenon which must be accounted for.

2) The Entropy Equals Zero Deduction; if the first law of thermodynamics is correct and the universe has net zero energy then net *entropy* must *also* be zero since zero energy *necessarily* entails zero entropy as implied by the *third* law of thermodynamics (because this law implies that net zero or infinite energy is a pre-requisite for net zero entropy). Given that energy in a closed universe is indeed net zero then it follows, given the third law, that net entropy too must be zero.

Consequently, it is possible to deduce (from the first and third laws of thermodynamics alone) that the second law of thermodynamics *cannot* logically be an inequality as currently expressed.

This deduction additionally suggests that *all* conservation laws – entropy, spin, electric charge, polarization, momentum etc. – are indirect expressions of energy conservation. In other words, if energy is net zero then all these others should logically be net zero too (which indeed seems to be the case).

to contract towards a point implying decreasing entropy. According to the Hawking-Bekenstein equation, though the universe *continues* to contract towards the Planck dimensions its entropy *cannot increase* and in fact must *decrease* as its *area* contracts. This is because the entropy of a black-hole is proportional to its area. This suggests that entropy tends towards zero as the universe collapses towards the Planck dimensions. This is inevitable unless the universe *ceases* to collapse after becoming a black-hole (the only other alternative), which the Hawking-Penrose calculations suggest it will not (otherwise, what is the status of its event horizon?)

There are therefore no clear reasons to believe that the last three minutes of a closed universe are anything other than rigorously symmetrical to the first three minutes – as a consequence of net entropy conservation. As such, the second law of thermodynamics should not be expressed as an inequality. Assuming a closed universe its proper form is;

$$\partial S = \frac{\partial q}{T} = 0.$$

8. Is Time Curved?

Hawking's initial objection to entropy conservation, that it implies time reversal, though incorrect (as was later conceded by him), is still prescient since entropy conservation *does* have implications for our current model of time. This model, let us remind ourselves, is rectilinear and therefore implicitly based on the Euclidean geometry utilized by Newton.

As we know however, the Euclidean model of space-time was overturned, in 1915, by Albert Einstein's Theory of General Relativity, which is firmly grounded in the curvilinear geometry of Bernhard Riemann et al. But the implications of this new space-time model, though correctly interpreted for space have never been carried over to our analysis of time as a curvilinear dimension. Why might this be?

The primary reasons for this state of affairs are that such an interpretation is counter-intuitive to an extra-ordinary degree and it also has no observable corollaries. Relativistic time curvature only becomes apparent on large scales and so is not detectable by means of observation. Whilst the hypothesis of curved space can be tested for in a variety of ways (this aspect of the theory was successfully tested for by Eddington as early as 1919), curved time is, in essence, a non-verifiable consequence of the same theory. Nevertheless, if space is indeed curved, as has been demonstrated, then ipso-facto time must be curved as well, even if no direct observations can confirm it. This is because relativistic time is part of a single geometry of space-time which is fully curvilinear at large scales. There cannot be an exception to this for any dimension, including time.

A further, technical, reason for this gap in our knowledge is that since the space-time manifold described by the general theory of relativity is, for reasons of convenience, most commonly expressed in terms of Minkowski space - a description which approximates to locally flat or *Euclidean* space-time – the radical implications of Einstein's theory for time are generally suppressed to students of the theory.

9. The Physical Significance of Curved Time.

Curved time has no meaning outside the context of a closed universe. This is because curved time implies *circularity* and hence closure as opposed to open ended expansiveness. This fact alone is a very strong argument in support of a "closed" model. Furthermore, since curvilinear time is a logical consequence of General Relativity it implies that the *prediction* of a closed universe is implicit in the General Theory of Relativity.

But what precisely does curved time mean in practice? If we imagine time as a circle traveling clockwise (depicted as the inner circle in the diagram below) it makes perfect sense to associate the maximal point of *spatial* expansion in our universe with the 90° point on the geodesic of time in our diagram. There is a clear logical symmetry in such an interpretation, as I think the diagram makes apparent;

The Cycle of Curvilinear Time.

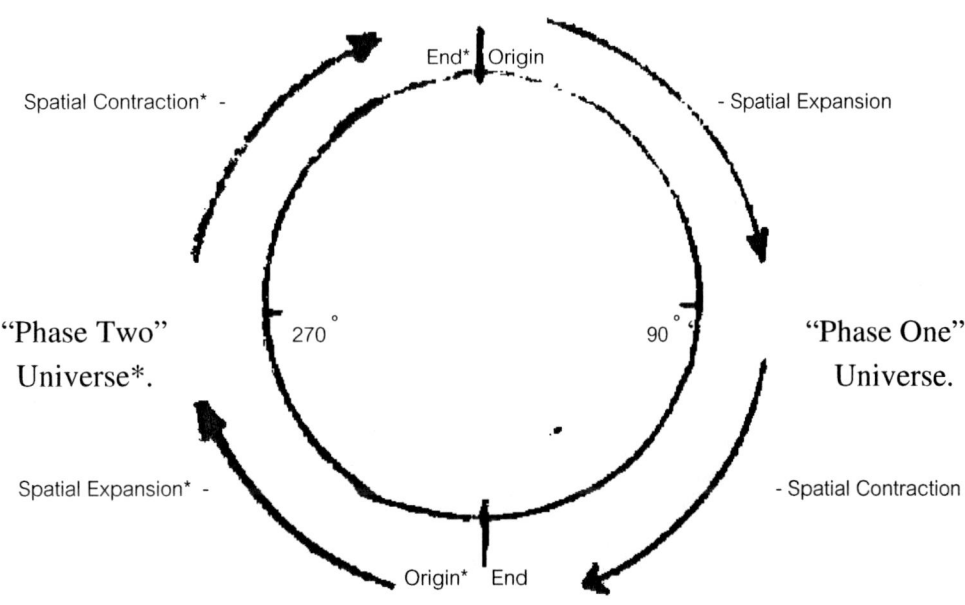

The only alternative to such an interpretation is to view the 180° point on the circumference of the time line as the point of maximal spatial expansion of our universe and the half-way point on the world-line of time. The only problem with this interpretation (which is the one I initially defaulted to many years ago) is that it does not account for the clear significance of the 90° point and the manifest change in declension that occurs thereafter. In my considered opinion this point has to have a marked physical significance (along the lines I have described) or else the "symmetry" between circular time and expanding and contracting space is lost.

If one accepts this line of reasoning however another difficulty emerges since it has our universe completing its *spatial* contraction (i.e. coming to an end) after only 180° - i.e. at the antipodal point of the geodesic of time – rather than the full 360° we might reasonably expect. This, if correct, begs the question; what happens to the other 180° of curvilinear time? And the logical answer is that a *second* or what I call in the diagram a "phase two" closed universe appears, mirroring the first.

10. Hyper-symmetry.

Why does the universe appear to be so profligate as to double itself in the manner just described? The brief answer is; because this facilitates the conservation of a variety of inter-related symmetries, notably; Charge Conjugation (the symmetry between matter and anti-matter which is clearly broken in our universe), Parity (space relational symmetry which is broken by various weak force interactions, notably beta-decay), C.P. (the combination of the prior two symmetries, which is spontaneously broken in weak force interactions) and Time (the breaking of which symmetry has recently been observed in the form of neutral kaon decay).

There is an additional symmetry involving a *combination* of all of the above (known as C.P.T. invariance) the breaking of which has never been observed. Such a violation would incidentally entail a violation of the Lorentz symmetry (which dictates that the laws of physics be the same for all observers under all transformations) upon the assumption of which the theories of relativity and the standard model of particle physics are founded.

Nevertheless, it is a strong possibility that C.P.T. (and hence the Lorentz symmetry) *are* spontaneously broken at the Planck scale where theories of quantum gravity become applicable. As a result of this suspicion (which I personally feel is entirely justified) theoretical physicists have suggested possible adjustments to the standard model (called standard model extensions) which allow for just such a possibility.[7]

Under a two phased model all of the above mentioned symmetry violations (those which have already been observed as well as those, such as C.P.T. itself, which are merely suspected) would naturally be repaired, leading (in accordance with Noether's theorem[8]) to the conservation of their associated quantities or quantum numbers. This is an outcome in accord with what I call the *second engineering function of the universe* (i.e. the *net* conservation of all symmetries) and is inconceivable under any other viable cosmological model.

The solution implied by most alternative cosmologies to the anomalies just outlined is simply to accept them as fixed features, unworthy (because incapable) of being explained. This is understandable on pragmatic grounds, but it is clearly unacceptable as a final position. As such these anomalies (concerning the missing symmetry of the universe) exist as hidden refutations of all cosmologies hitherto that *exclude* the notion of hyper-symmetry.

After all, if super-symmetry and multi-dimensional string theory can legitimately be postulated on the grounds of the otherwise incurable problems they solve in the standard model, then how much more so can hyper-symmetry and the two phased universe, whose purpose is to facilitate the conservation of a vital set of symmetries that are otherwise wholly unaccounted for? The hypothesis is, after all, not an artificial one, but springs naturally out of a consideration of the properties of curved time in the context of Einstein's General Theory of Relativity. This is more than can be said even for super-string theory and its variants.

Another important conservation facilitated by hyper-symmetry (i.e. by a two phased universe) is that of baryon number. Baryonic asymmetry (a key feature of our universe) is a by-product of spontaneous symmetry breaking, particularly relating to charge conjugation. The net conservation of baryonic symmetry and of net baryon number of the *two* phased (or hyper-symmetrical) universe serves to account for the *asymmetry* we observe in our current universe. What this translates to in practice is that whilst *quarks* predominate in our current phase, *anti*-quarks (to use the conventional terminology, which is

---

[7] C.P.T. and Lorentz Symmetry II, Alan Kostelecky ed, (World Scientific, Singapore 2002).
Lorentz-Violating extension of the standard model. Dan Colladay and Alan Kostelecky. Phy. Rev. D58, 116002 (1998).
[8] Noether's theorem is informally expressible as; to every symmetry there corresponds a conserved quantity and vice versa. For example, the Lorentz symmetry corresponds to the conservation of energy. For a more formal account of this theorem see H.A. Kastrup, Symmetries in Physics (1600-1980) (Barcelona, 1987), P113-163.

obviously biased) are bound to predominate in the same way in the next phase (a phase we might term the "*anti-universe*", but it is of course all relative). This tells us that the net baryon number of the two phased universe as a whole *must* be fixed at zero. This is only deducible when hyper-symmetry is taken into account, as it must be. What is more, these *same* arguments can be applied to the analysis of apparent lepton asymmetry as well and to net lepton number conservation in an hyper-symmetrical universe.

It also follows from the above observations concerning baryons that the three so called "Sakharov conditions" deemed to be pre-requisites for baryogenesis to take place (i.e. thermal disequilibrium, C.P. invariance violation and baryon number violation), all of which have been empirically observed, are fully admissible in a hyper-symmetric cosmology which possesses the added benefit of rendering these conditions conformal rather than anomalous as is currently assumed to be the case. In other words, the new cosmological paradigm of hyper-symmetry allows for the Sakharov conditions to obtain *without* net symmetry violations of *any* sort truly occurring. Which is a really remarkable result and not achievable under any other known paradigm.

In addition to this, there is no longer any need to hypothesize an "inflationary period" to "blow away" (read "sweep under the carpet") the apparent initial asymmetries (possibly including that of C.P.T. itself)[9] which are all fully accounted for under this paradigm.

My suspicion however, for what it is worth, is that C.P. violation may have more relevance to the issue of *leptogenesis* than it does to baryogenesis since G.U.T. scale or strong force interactions show no evidence of C.P violation, though they may reveal evidence of charge conjugation invariance violation. This suggests that charge conjugation invariance violation and *not* C.P. invariance violation is the *true* precondition for baryogenesis. C.P violation is however a necessary precondition for leptogenesis, as is thermal disequilibrium and lepton number violation.

It is also important to note that Parity violation in an *anti-matter* universe is the mirror opposite to parity violation in a *matter* universe such as our own. As such, the combination C.P. is indisputably *conserved* in a two phased universe, as is its associated quantum number. This is obviously what we would expect from hyper-symmetry, which alone gives us this key result.

The final missing symmetry – that pertaining to time – is also unquestionably accounted for by what I call "hyper-symmetry" in a two phased universe, but the argument for this is a little less self evident. It is perhaps useful in this regard to imagine neutral kaon decay (our best serving example for time asymmetry) as occurring (so to speak) in a direction *opposite* to that of the familiar arrow of time. With such a model in mind net time symmetry conservation becomes easier to understand with reference to the diagram presented in the previous section. This is because it becomes evident that time symmetry violation means precisely *opposite* things for each separate phase of the two phased universe. This is due to the fact that the arrow of time is metaphorically moving Southwards in phase one (implying north pointing symmetry violations) and Northwards in phase two (implying south pointing symmetry violations). In consequence it is easy to deduce that net time symmetry (which is clearly broken in our one phased universe) is, in actuality, *conserved* in the context of our hyper-symmetric or two phased universe. This is especially satisfying since if time were *not* conserved (as is the de-facto case in *all* other cosmologies) it would seem to entail a violation of the conservation of energy as well.

And this complete set of symmetries effortlessly and elegantly provided by our hyper-invariant cosmology (itself the logical consequence of curved time and hence of general relativity) *cannot* be achieved in any other viable and convincing way, a fact that potential critics of the new paradigm might wish to take note of.

---

[9] C.P.T. Violation and Baryogenesis. O. Bertolami et al. Phys, Lett. B395, 178 (1997).

11. The Engineering Functions of the Universe.

The two phases of the universe are thus indicative of an elegant efficiency which serves to effect the **net conservation of all symmetries**, which is a principle that describes what I call the *second* engineering function of the universe. This principle is itself complementary to the *primary* engineering function of the universe, which is, (as discussed in detail earlier), the avoidance of the (logically absurd) zero or infinite energy equilibrium represented by the radiation origin (where all symmetries are ipso-facto conserved). The universe therefore represents the "long way round approach" to conserving symmetry, an approach *forced* upon it by the fact that nature abhors vacuums and infinities alike.

The universe, in other words, cannot take the *direct* approach to the maintenance of equilibrium or perfect symmetry since this would leave it in violation of the *primary* engineering function. This double bind (in which the universe must conserve *all* symmetries but *cannot* avail itself of zero or infinite energy), is the underlying raison d'etre for both the *form* as well as the very *existence* of our universe.

For those who point to the apparent outlandishness of this hyper-invariant cosmology I can only make the following defence; General Relativity necessitates the hypothesis of curvilinear time and this model is the only viable *interpretation* of curvilinear time. What is more, the model appears to clear up a number of outstanding problems (concerning obscure but important symmetry conservations) that are not otherwise soluble. It is these considerations which lead me to conclude that this solution is not merely elegant and efficient but that it is also true.

12. Pierre Curie and the Principle of Sufficient Reason.

Perfect symmetry describes a condition of either zero or infinite energy because perfect order implies zero entropy which in turn implies one or other of these two cases, both of which (crucially) are mathematical idealities.

Of obvious relevance at this juncture is Pierre Curie's observation that it is *asymmetry* that is responsible for the existence of phenomena.[10] In other words, *all phenomena are the product of symmetry violations*. This principle implies that the universe itself must be the product of spontaneous symmetry breaking in the state of perfect order just defined. Indeed, it amounts to nothing short of a proof of this.

Leibniz's famous principle of sufficient reason however suggests that a *reason* for the deviation from equilibrium initial conditions must be adduced, otherwise perfect symmetry could never be broken. It is this reason which is supplied by the first engineering function, itself a logical derivation from the third law of thermodynamics (see section twenty nine for the details of this derivation). Thus, the first engineering function is isolatable as the *reason* for the existence of the universe, whilst the second engineering function (the principle of net invariance) is primarily responsible for the universe's peculiar (hyper-symmetrical) *form*. Hence the centrality of these two hitherto un-postulated principles of inductive philosophy.

The implication of the first engineering function therefore is positive; *there must be existence* (and hence spontaneous symmetry breaking to bring it about). The implication of the second engineering function however apparently countermands this; *there must be no existence* (i.e. all symmetries must be preserved since this is a rational requirement for the equilibrium of the universe).

The compromise that is reached as a result of this impasse is what I have called *Universal Relativity*, or eternal transience, which is indeed what we *do* observe. In other words, there exists neither *substantial being* (in line with the second engineering function) nor yet *complete nothingness* (in line with the first). An effortless and, in some respects, *uncanny* synthesis (of being and nothingness) is what results from this.

Furthermore, it is precisely the *lack* of these two principles (principles which happen to incorporate the Lorentz symmetry) which accounts for the perceived inadequacy of classical empiricism in accounting for inter-subjectivity, prior to Lorentz and Einstein. Since these two principles are at once apriori *and* empirically justified then this points to limitations (as Kant has also demonstrated) to the basic assumption of Empiricism; that all concepts must be aposteriori.

Infact, the Lorentz symmetry alone is sufficient to account for the phenomenon of objectivity, which Kant's philosophy had (according to Kant himself) been constructed in order to explain. It therefore serves to confirm and lend greater precision to certain aspects of Kant's analysis.[11]

Furthermore, the postulation of the *first* engineering function (as a logical derivative of the third law of thermodynamics) allows the principle of invariance to cohere alongside the principle of sufficient reason without a necessary contradiction, which is not otherwise possible.[12] In other words, although primal symmetry breaking *is* spontaneous (from a causal perspective) it is *not* therefore without sufficient

---

[10] Pierre Curie; Sur la symmetrie dans les phenomenes physiques. Journal de physique. 3rd series. Volume 3. p 393-417.

[11] Kant argued that the ambit of pure reason is limited to an empirical subject matter and that where apriori reasoning strays beyond this subject matter it falls into nonsense and self-contradiction (i.e. the antinomies of reason). Yet Kant's system implicitly adopts a *metaphysical* point of view from which to state these things and so (rather like Wittgenstein's *Tractatus*) falls foul of its own strictures.

The idea behind logical empiricism was (in my opinion) to avoid this self-contradictory dependence inherent in Kant's system. Unfortunately Frege's project (which evolved into *logical positivism*) fell foul of a different dependence, the dependence on a limited *classical* conception of logic. At present, prior to *Principia Empirica*, nothing has succeeded in replacing a now discredited logical positivism.

[12] At any rate the elementary *fact* of spontaneous symmetry breaking means that we must either *abandon* the principle of sufficient reason *or* we must make use of the *first engineering function* as I have postulated it. In either case an important advance is made.

reason (it was Kant's mistake to identify sufficient reason with the principle of causality). It is simply that the "sufficient reason" (in this case, the first engineering function) is acausal and hence apriori in nature.

Kant's error, (as I see it) has been replicated more recently in the work of Donald Davidson[13] where he argues that a *reason* is ipso-facto a *cause*. It is clear that in an indeterminate universe (that is, a universe where *indeterminism* forms a substrate to causality, rendering causality itself, i.e. *determinism*, a *contingent* or special case of *indeterminism*) spontaneous symmetry breaking *cannot* have a cause. To have a cause implies the existence of *phenomena* acting as causal agents. As Curie's remarkable principle indicates however there can be *no* phenomena independent of symmetry breaking, ergo the *reason* for phenomena cannot be their *cause*. Rationality is thus a more general concept than causality and retains the latter as a special case.

As I have indicated the *reason* for phenomena lies in the realm of logic and symmetry (notably the apriori-like nature of what I have identified as the primary engineering function) and *not* of causality. Ergo reason and causality are *not* identical, contra Davidson and Kant. (This view might more formally be expressed as; *reasons only become causes at the classical limit.*) The fact that Davidson (rather like W.V.O. Quine in other instances) ignores the role of apriori or *logical* (i.e. acausal) reasons indicates the empiricist bias of the school of philsoophy from which his and Quine's work springs. This confusion of reason and cause is, I would say, a fatal lacuna of Empiricism.[14]

It also follows from this that rational explanation is *not* dependent on the universality of causality and determinism, but is more general in nature. This is because "sufficient reason" can be imagined *independent* of these embarrassingly limited principles, as I have just shown.

This delinking of Rationalism from determinism forms a major argument in support of my suggested renovation of the Rationalistic (and, more fundamentally still, the *Logicist*) program argued for later in this work. It also refutes the basic premise of Kantianism which is that apriori knowledge independent of aposteriori experience is impossible. In so much as Rationalism and Empiricism or Determinism *are* linked, Empiricism is the junior partner.

It is therefore ultimately *Rationalism* and *Logicism* which I interpret as underpinning the viability of empiricism. They, as it were, *encapsulate* empiricism in a still broader philosophical framework that is both Rationalistic *and* Logicistic in its presuppositions. And in this sense, in agreement with Kant, I do not see Rationalism and Empiricism as fundamentally antagonistic. However (unlike him), I *do* see Rationalism as possessing an ambit demonstrably wider than the purely empirical, which I will explain and illustrate in various ways as we go on.

In a very real sense therefore Rationalism and Empiricism are *not* coequal as Kant supposed, but Rationalism *incorporates* empiricism as a *part* of its overall domain, a domain which, in *its* turn, is fully circumscribed by logical analysis. However, it will be found that neither the Rationalism *nor* the Logicism reffered to in this new schema (which purports to supply the logical and rational foundations for empiricism itself) is or can be of the *classical* variety traditionally associated with these two failed (because hitherto based on *classical* assumptions) philosophical doctrines. More of this later.

In the case of our universe then "sufficient reason" for its very existence is supplied by the first engineering function which forbids the manifestation of zero or infinite energy (in effect forbidding perfectly symmetrical and hence equilibrial) conditions. Quantum mechanics (which notoriously *allows*

---

[13] *Actions, Reasons and Causes. Journal of Philosophy*, 60, 1963. Davidson's argument appears to have the effect of making rationality dependent on causality, which is an unsustainable view as my work hopefully demonstrates. The narrowly empiricist bias implicit in such an argument is also worthy of note.

[14] My own view is that Curie's principle also serves to *prove* that the universe must have an origination in a state of perfect order as described in this work. This is because *all* phenomena (according to the principle) require *symmetry to be broken* in order to exist. This presumably includes phenomena such as *quantum foam*. Thus Curie's principle stands in direct and stark opposition to the current standard model for the origin of the universe; Alan Guth's *Inflationary theory*.

This theory posits an origin for the universe out of the so called quantum foam, which suggests that the foam is an eternal and uncreated phenomenon, something *forbidden* by Curie's principle. The only alternative to Guth's theory is a creation *ex nihilo*, which is to say creation out of a state of *perfect order*, due to spontaneous symmetry breaking. This indeed is the only *logical* explanation that is *also* empirically viable, as I have already shown.

For a more detailed refutation of Guth's alternative see section fourteen.

for spontaneous symmetry breaking) is merely the empirical *mechanism* or medium through which these various logical and physical principles are *necessarily* enacted, creating the universe in the process.

13. Open Or Closed?

Current observations vary on an almost monthly basis as to whether the universe has enough mass-energy to be closed. Attaching oneself to one view or another is therefore a little like joining a political party. The best piece of observational evidence (based on recent readings from the cosmic background radiation[15]) points strongly to a flat solution however. This seems to rule out the "closed" solution required by my preceding model, except that, in reality, a flat and a closed solution are not as mutually exclusive as they might at first sight appear to be.

The point to make clear is that observation alone can *never* solve this conundrum *if* the flat solution is, as it is widely believed to be, the correct solution. This is because in such a case the results of observations will always be too close to favouring *all three* solutions to be of decisive significance. Such observations will inevitably fall victim, (in what amounts to a cosmological equivalent of the uncertainty principle), to whatever margin of error happens to be built into them. And, in the case of observations based on deep space astronomy, these margins of error are always likely to be very large indeed. Unfortunately there is not enough recognition, (or indeed any that I am aware of), of this latent epistemological problem within the cosmological community, who continue sedulously counting their "w.i.m.ps" and baryons as if no such difficulty existed at all.

Paradoxically then, the closer observation takes us to a flat solution, the further away from certainty we become. Clearly this is an intolerable circumstance from the point of view of resolving this issue, but what other alternatives exist?

Fortunately a way out of this dilemma can, I believe, be supplied by means of recourse to theoretical considerations alone. It is with reference to theoretical considerations for example that we are able, I believe, to rule out the flat solution entirely. This is because the flat solution is a *classical* one and therefore negated by quantum theory. The reasons for this are two-fold. Firstly, the "critical density" determining the fate of the universe is so precisely defined (at 10^{-29} grms per cm^3) that any fluctuation at or around this level, of the sort dictated by quantum theory, is liable to push the density over this critical limit and into a closed situation.

The second indication that $\Omega = 1$ is a classical solution is that it points to a universe which expands eternally but at a *diminishing* rate. This is plainly absurd from a physical point of view since the irreducible quality of the Planck dimensions must ultimately act as a break on this transcendental process. As a *classical* solution therefore the flat universe seems to fall victim to a cosmological application of the law of excluded middle, as enacted by the mechanisms of quantum theory. All that remains therefore is for the cosmological community, particularly the *relativistic* cosmologists, to accept the implications of this fact.

Having ruled out the flat solution with recourse to theory alone, is it possible to co-opt pure theory so as to select between the two remaining options? I believe that there are two very strong such reasons, (supplied by relativistic cosmology and by quantum cosmology respectively), for selecting in favour of the closed solution. The first has already been mentioned in connection with the hypothesis of hyper-symmetry and curvilinear time, which is an hypothesis that only makes sense within the context of a closed universe as we have seen.

The second "proof" is somewhat more elaborate but, if anything, even more compelling. Since, (as was discussed in sections three and six), *all* energy in the universe is ultimately a form of *vacuum energy* (including the so called "dark-energy") and since *quantum theory* requires that *all* vacuum energy ultimately be conserved, (in order to avoid contravening the first law of thermodynamics), it therefore follows that the universe must be closed since the conservation of (ultimately *zero*) energy *can* only occur in a closed universe. Ergo the universe *is* closed, irrespective of what current observations *appear*

---

[15] P. de Bernardis et al, 2000. Nature 404 955.
S. Padin et al, 2001. Astrophys. J. Lett. 549 L1.

to be telling us. Incidentally, this argument can also be adapted to prove that the mysterious "cosmological constant" (lambda) must also be net zero, since lambda is inescapably identified with vacuum energy (which is certainly net zero). This is logically unavoidable, but should not be unexpected in a universe where all symmetries are ultimately conserved (*at* net zero) for "engineering" reasons.

What is more, the link between the cosmological constant and vacuum energy (i.e. between relativity and quantum theory) points tellingly to the central role of vacuum energy in fuelling the expansion of the universe, since it is lambda (Einstein's "greatest mistake") which ultimately defines whether the universe is stationary or expanding according to General Relativity.

The reliability of recent observations (based on the ad-hoc "standard candle" method of gauging interstellar distances) which suggest that the universe is expanding at an *accelerating* rate and that lambda is therefore non-zero, can certainly be challenged, based as they are on the *least* reliable form of empirical observation (deep space astronomy) but the first law of thermodynamics, which dictates that vacuum energy and hence lambda be net zero, cannot realistically be violated. However, there is nothing in the laws of thermodynamics which in principle prevents the cosmological constant from (paradoxically) *fluctuating* around a net zero value. Another possibility is that the dark energy fuelling the apparently accelerating expansion eventually exhausts itself and becomes negative, thereby conforming to the first law of thermodynamics and ultimately contributing to a gravitational collapse. Given the growing strength of the evidence for an accelerating expansion this is the possible explanation that I find most plausible at present.

Whatever the case may be concerning these diverse matters it remains a fact that no universe unless it is closed can avoid violating the first law of thermodynamics. Therefore on purely *theoretical* grounds alone it seems reasonable to assume that the universe *is* closed, even if recent empirical evidence is ambiguous on the matter.

14. The Critique of Pure Inflation.

According to the hot big-bang model briefly outlined in section one, the absence of significant gravitational constraints prior to the "birth" of matter after 10^{-36} seconds A.T.B. *must* translate into a rapid expansion of the universe up to that point. There would be no "anti-gravitational" requirement (of the sort that still troubles cosmologists), *prior* to 10^{-36} seconds since there would have been very little gravity needing counteraction. Baryogenesis (the generation of quarks) would naturally be triggered only *after* this point by the cooling, i.e. entropy, of the high energy radiation released by primal symmetry breaking.

The expansion of the universe is therefore triggered at 10^{-43} seconds by this "spontaneous symmetry breaking" (when gravity separates off from the other forces) and is fuelled by the enormous amounts of vacuum energy liberated by this decay of a perfect state of order. Symmetry breaking in zero entropy conditions (i.e. the radiation origin) provides unlimited potential energy which translates directly into a very rapid and orderly spatial expansion or heat flow. It is this event alone which is capable of triggering *some* form of rapid expansion of space time. No other hypothesis is either necessary or appropriate.

Although a trigger for exponential expansion at 10^{-36} seconds A.T.B. (when the strong force separated from the others) *is* conceivable (and is the favoured choice amongst proponents of the so called "inflationary theory") it is not obvious why *this* phase transition and not the immediately prior one (at 10^{-43} seconds) should have been the trigger for an expansion. Indeed, because the separation of the strong force is ultimately associated with baryogenesis (the creation of matter) it seems *more* probable that G.U.T. symmetry breaking at 10^{-36} seconds is commensurate with a *deceleration* in an already supercharged rate of expansion rather than with "inflation". 10^{-36} seconds represents the point at which energy from the big bang begins its damascene conversion into matter, but the *origin* of that energy and hence of the expansion of the universe must lie with spontaneous symmetry breaking at 10^{-43} seconds. The negative pressure of the *actual* vacuum (devoid of gravitational constraints) ensures this expansion out of the quantum origin.

Since the heat-flow of the universe is *directly proportional to its area* it follows that the *expansion* of the universe *must* begin at its origin since this heat-flow by definition *also* begins at the radiation origin. The proof of this hypothesis lies in an adaptation of the Bekenstein-Hawking area Law. This law in its *original* form links the entropy of a black-hole (S) to the size of its area squared (A^2). But an equal case can be made for linking the entropy of the universe to *its* size as well;

$$S_{uni} = C_n \times A_{uni}^2$$

For the universe *not* to have expanded out of the quantum origin in the manner suggested therefore amounts to saying that entropy (heat-flow) comes to a (temporary) halt at this juncture only to resume again after 10^{-36} seconds A.T.B. which is plainly impossible.

There can be little doubt therefore that the universe *was* expanding out of the quantum origin. Indeed the quantum origin represents a type of expansion in its own right. The universe, in other words, could be said to have expanded in diameter from $0 \rightarrow 10^{-33}$ cm in 10^{-43} seconds flat – which is a rate of expansion exactly equal to the velocity of light. In this case the *radiation origin* can be seen as the *true* source of the expansion and the quantum origin as the first of many phase transitions to exert a *braking* effect on this extraordinary rate of expansion. And it is the Planck mass (10^{-5} kilos) which exerts this initial brake on the velocity of the expansion coming out of the quantum origin. Baryogenesis exerts the *next* brake after 10^{-36} seconds A.T.B., leaving no room for an exponential expansion of the sort proposed by Professor Alan Guth. All that an expansion that originates at light velocity *can* do is decelerate with each successive phase transition. The G.U.T. symmetry breaking transition after 10^{-36} seconds A.T.B. is no exception to this rule.

If the supporters of Guth's model of exponential expansion wish to place the onset of so called inflation *later* than 10^{-43} seconds, they therefore have to explain *why* they think the universe would not expand at a phenomenal rate as a result of the spontaneous breaking of symmetry represented by this point. Because there is *no doubt* that it did. This is because the quantum origin represents a disequilibrial consequence of *already* broken symmetry between the fundamental forces, one which results in the release of tremendous amounts of potential energy. And this energy can flow nowhere except in the form of an extraordinarily rapid expansion. This is due to the fact that quantum theory forbids further contraction below the Planck dimensions whereas disequilibrium prevents stasis. Ergo, the big-bang expansion out of the quantum origin is as *inevitable* as is our model for the quantum origin itself, upon which it logically depends. There simply *is* no viable alternative to this overall picture.

Incidentally, the homogeneity observed in the cosmic background radiation (the so called "horizon problem") is also necessarily explained by this model. This is because the nearly perfect state of symmetry represented by the quantum origin translates *directly* into an highly *orderly* expansion of the universe which in turn *ensures* that high (though not perfect) degrees of homogeneity in the cosmic background (i.e. *black-body*) radiation are conserved, even in areas of space that have been out of communication with each other since T=0. That the cosmological model presented here is true (including and especially our interpretation of the radiation origin) is confirmed by the fact that the cosmic background radiation displays a *near perfect* black-body spectrum. And this is indeed an *astonishing* confirmation of what is an highly elegant picture.

The question therefore is as to the *nature* of the expansion after 10^{-43} seconds and not the *timing* of its initiation. The somewhat cumbersome mechanisms Guth suggests (the so called "inflaton field" and the "false vacuum") are rendered redundant by the already very rapid (but not exponential) expansion of the universe prior to this point. It is, in any case, the separation of the *gravitational* and not the mythical "inflaton" field which triggers the putative expansion and only some future theory of quantum gravity is therefore likely to shed more light on this process.

The time between 10^{-43} seconds and 10^{-36} seconds can clearly be seen as an era of unhindered spatial expansion according to the hot big-bang model. To say that this expansion was exponential in nature is difficult to justify however, given the clear facts surrounding the quantum origin mentioned above. The best that *can* be said, with confidence, is that the conditions (i.e. the absence of significant amounts of matter prior to baryogenesis) *were* almost perfect for a de Sitter type expansion (i.e. for what Guth unhelpfully terms a "false vacuum"). These conditions indeed, which lasted for 10^{-7} seconds, render Einstein's early objection to the de Sitter solution redundant. And this alone is a remarkable and intriguing fact.

A natural termination to a de Sitter type expansion is also supplied, sometime after 10^{-36} seconds, by baryogenesis and the attendant emergence of gravity as a significant countervailing force. In any case, even though the initial expansion was probably *not* of the de Sitter type, it would inevitably have experienced deceleration as a by-product of the phase-transition at this juncture. Of this there can be little doubt.

The idea of an exponentially expanding space-time continuum in this brief era *is* a compelling one (and Guth deserves credit for at least reviving interest in it). It is compelling though, *not* because of the various cosmological problems it is alleged to solve, but purely and simply because of the de Sitter solutions to the equations of general relativity. However, the fact that a model has great explanatory power does not itself make it true and inflationary theory is unquestionably a great example of this fact. The aforementioned cosmological problems, for example, can all just as well be solved assuming *any* model of a sufficiently expanding universe. The exceptions to this are the previously discussed "horizon problem" and the monopole production problem, which is *so* extreme that it almost certainly points to flaws in the current "standard model" of particle physics, whose ad-hoc nature is well known to its adepts. The *true* solution to this problem therefore has very little to do with cosmology, inflationary or otherwise.

Guth's explanation of the so called "flatness problem" (the *apparent* zero curvature of space) also feels inadequate and could equally well be used to justify *any* model of a sufficiently expanding universe. A more *fundamental* explanation as to *why* the initial conditions of the universe might be so finely tuned to a flat solution is that the impulsion to conserve *all* possible symmetries (what I elsewhere call the "second engineering function of the universe") necessitates the natural selection of what amounts to a perfectly balanced, near equilibrium solution to the geometry of the universe.

But even if Guth's explanation of the "flatness problem" is the correct one (in spite of being comparatively superficial) there is no reason to assume that it is necessarily exclusive to (and hence confirmatory of) an inflationary model. The same point can be made concerning the "quantum" explanation of the ultimate origin of the large-scale structure (as distinct from homogeneity) of our universe. This explanation (which is undoubtedly correct) has absolutely no inevitable association with the inflationary model and so cannot be arbitrarily interpreted as representing a confirmation of that model. It will work just as well with almost any finely tuned model of a rapidly expanding universe.

If, as I suggest is indeed the case, compatibility cannot be discovered with spontaneous symmetry breaking at 10^{-43} seconds, then, irrespective of the above considerations, the de Sitter solution (in *any* form) must be abandoned as an incorrect cosmological model. To place it at a later date (as Guth et al do) simply does not work because, as mentioned before, the universe is *already* expanding at a ferocious rate by the time Guthian inflation is meant to start! This leaves *no room* for an exponential expansion, which might even transgress the light velocity barrier imposed by the Special Theory of Relativity. It also places unbearable stress on the theoretical structures that have to explain how and when this process is meant to start. Guth's "late start" is therefore incompatible with a correct understanding of the hot big-bang model. It is a non-starter.[16]

But another equally compelling set of reasons exist that argue against a later period for expansion, "inflationary" or otherwise. Guth's model implicitly and explicitly postulates a period of chaos and disorder in the era *preceding* alleged inflation, an era which effectively plugs the gap between the Planck time and the onset of inflation. By placing the origin of the expansion in the wrong place Guth in effect *creates* this chaos and disorder.

It is out of *this* chaotic period that the orderly bubble of inflation (the so called "false vacuum") is supposed to have taken hold as the product of a quantum fluctuation. Indeed, "inflation" seems to exist as some kind of latter day "epicycle" designed to justify the maintenance of this irrational picture. But as we have seen from our analyses hitherto the period under discussion was very far from being one of high entropy. Indeed the further back in time we go, the *more order* we find. And this is a *necessary* consequence of Curie's principle of symmetry, as discussed earlier.

It simply is not necessary to give the clock of the universe (so to speak) the artificial winding up that Guth's model requires. In any case, the universe *could not* be the product of a statistical fluctuation out of chaos because high entropy states simply do not possess sufficient potential energy to do the job. Appeals to the randomness of quantum mechanics are not enough to overcome this key objection as Guth seems to think. As such, expansion must be the product of spontaneous symmetry breaking within a state of perfect order (i.e. one replete with *unlimited* supplies of potential energy) instead. This indeed is the *only* viable alternative to Guth's flawed model.

For those who demand to know the source of such a state of perfect order it will hopefully suffice to say that perfect order is *not* a created state, but rather describes a situation of zero entropy. Nothing is therefore required to explain the origins of a state which is definable only in terms of what it is not. And this, incidentally, is a very strong argument in support of ontological nihilism, since it is this

---

[16] The fact that there are so many divergent versions of the inflationary theory is also an argument against it. Thomas Kuhn has written of such situations as being indicative of a crisis in the underlying theory; "By the time Lavoisier began his experiment on airs in the early 1770's, there were almost as many versions of the theory as there were pneumatic chemists. That proliferation of a theory is a very usual symptom of crisis. In his preface, Copernicus spoke of it as well."
Thomas Kuhn, *The Structure of Scientific Revolutions,* p71, section 7. Univerity of Chicago Press. 1963.

indeterminate "non-state" which lies at the origin and essence of our mysterious and beautiful universe. As such, it is surely more relevant to ask the supporters of the Guth paradigm – where did all the entropy come from?

A final argument concerning the insufficiency of the inflationary model pertains to its account of sundry asymmetries relating to charge conjugation, parity and time (see section ten for a more complete discussion of these). These asymmetries are allegedly "blown away" by exponential inflation, i.e. they are balanced out by conveniently equal and opposite activity in other (again conveniently) invisible (that is, out of communication) regions of the universe. This account (which amounts to sweeping the problem under the carpet) merely serves to highlight the profoundly *unsymmetrical* nature of Guth's paradigm rather than properly accounting for the asymmetries which, in my view, require *hyper*-symmetry (i.e. more not less symmetry!) in order to *genuinely* explain them.

Inflationary theory, which teeters on the brink of metaphysics, is in consequence, as may have been gathered by the attentive reader, intensely difficult to falsify and yet, given the correct application of fundamental physics it can, as I have hopefully shown, be seen to be insupportable and incompatible with the facts. And the essence of this falsification lies in demonstrating that (in the light of fundamental theory) *any* form of exponential expansion is both impossible and unnecessary.

15. Net Perfect Order.

Net zero entropy (as argued for in section seven et-seq) necessarily equates to net perfect order. This inescapable and seemingly mystical conclusion is not of course the consequence of transcendental design of some sort but is simply the by-product of various conservation laws and hence, ultimately, of nature's underlying parsimony.

The key distinction between the *ideal* perfect order of the radiation origin and the *real* perfect order of the universe as a gestalt entity lies, of course, in the use of the word "net". Not at any time is the universe free from entropy or disorder (if it were it would be free of "being" as well, thus effectively violating the first or primary engineering function of the universe) and yet, due to the conservation of all symmetries (the second engineering function) the universe as a whole (through time as well as space) must be interpreted as an example of *net* perfect order. In other words, due to the influence of the two engineering functions acting in concert together, we ipso-facto live in a universe of *dynamic* or real equilibrium, mirroring (or conserving) the *static* or ideal equilibrium of the radiation origin.

Recognition of these facts helps us to address the perennial mystery of why the physical constants take on the apparently arbitrary yet very precise values that they do. These values, we may remember, can only be ascertained directly, by means of empirical measurement and cannot, for the most part, be deduced mathematically from fundamental theory. I submit that this state of affairs will not be altered one wit should a future so called "final theory" incorporating quantum gravity someday emerge.

It seems to me that acknowledgement of the validity of what I term the "second engineering function of the universe" – i.e. the net conservation of perfect symmetry – allows us to understand that the ensemble of the physical constants (and in particular the fundamental constants) acting in concert together, are what make the second engineering function fulfillable in the first instance.

From this we may surmise that the *values* taken by these numerous constants (irrespective of whether they are all permanently fixed or not) are "naturally selected" at the radiation origin so as to allow this key engineering function to fulfill itself in the most efficient (i.e. parsimonious) manner possible. As a result of which (since they are *self organizing*) these values cannot be derived from theory apriori (except in a piecemeal fashion) but can only be measured a posteriori.

That nature is *obliged* to obey all conservation laws is also the ultimate source of the global or "classical" order underlying local "quantum" unpredictability. The great conservation laws in a sense *guide* (or at least *constrict*) quantum indeterminacy without removing it. Their existence strongly implies that quantum behaviour is inherently indeterminate *without* being random. And, in much the same way, the requirements of the conservation laws ensure that the values ascribed to the constants of nature are *not* random (i.e. are similarly constricted) even though they *cannot* be derived from a fundamental theory.

It is the great conservation laws therefore that are ultimately responsible for the rationality we appear to detect across nature and even within ourselves,[17] notwithstanding rampant indeterminacy. Yet it is ultimately a nihilistic or "zero sum" rationality.

---

[17] This may remind us of Kant's famous peroration at the end of the second Critique; "The star strewn heavens above me and the moral law within me."
Principia Logica in many respects represents an attempted renovation of Kant's system (or at least of some of its insights) in the light of discoveries in modern logic and physics. These, I feel, give a modern rational philosophy a firmer objective base, precisely of the sort Kant aspired to, whilst retaining the *essence* of his critique of the limitations of both Empiricism and Rationalism. Some of his metaphysical concepts and distinctions are however abandoned as imprecise and unnecessary.

16. The Physical Significance of Net Perfect Order.

It is hopefully not too fanciful to suggest that net perfect order is the origin of much of the cyclicality we observe everywhere in nature. Such fractal cyclicality leads naturally to the hypothesis that the universe *itself* is cyclical. This hypothesis is clearly a feature of the cosmology presented here, but has its historical roots in Richard Tolman's cyclical model of the 1930s, a model postulated in the wake of Hubble's announcement, a year earlier, that the universe is in fact expanding. This model, though revived in the 1970s by another American cosmologist, Robert Dicke, was displaced in the 1980s by Alan Guth's "Inflationary cosmology" which has remained the dominant paradigm ever since. Recent astronomical observations of an apparently accelerating expansion have however led to a renewed questioning of this dominance and to a concomitant revival of interest in the cyclical model.

There are two primary objections to the cyclical or "oscillating universe" paradigm which have led to its historical displacement and both these objections have already been countered in this work. The first objection – that increasing entropy will eventually cause the cycles of expansion and contraction of space to "wear out" and exhaust themselves – is based entirely on the assumption of entropy non-conservation which was challenged in section seven. In essence I argue that heat-flow must necessarily *decrease* in a contracting universe, thus leading to net entropy (i.e. heat-flow) conservation in any given cycle. Granted entropy conservation, cosmological cyclicality of the sort postulated by Tolman, Dicke and myself can indeed be eternal. Such a solution also has the happy side-effect of rendering the recent cumbersome and highly inelegant Steinhardt-Turok variant of the cyclical model superfluous.

The second objection to the cyclical paradigm; the singularity problem; has already been dealt with in our treatment of the quantum origin. A universe that contracts to the Planck scale, as we have seen, is obliged by quantum theory to bounce into a renewed cycle of expansion, effectively obviating the classical problem of the singularity. There is therefore nothing, in principle, to prevent the universe from flipping eternally from quantum origin to quantum origin. These origins, in fact, would seem to constitute the twin pivots around which our universe diurnally turns from cycle to cycle.

This interpretation is backed up by a consideration of the little known third law of thermodynamics (also called "Nernst's heat theorem") which postulates that energy can never be entirely removed from a system (such as the universe itself for example). This law can be seen to express itself, in this context, in the form of the fundamental character of the Planck dimensions. Thus, once again, the quantum format of the universe is demonstrated to be an almost inevitable consequence of what I call the first engineering function of the universe; i.e. ***the avoidance of zero or infinite energy conditions***. The deep dependence of *all three* laws of thermodynamics (and hence of quantum physics itself) on the two engineering functions of the universe points to the ultimate ontological significance of these two functions. They constitute nothing less than the apriori foundations of the universe itself and of all the laws that are deemed to govern it. Their postulation is thus a vital part of any comprehensive rational philosophy of the sort proposed in this book.

The third law of thermodynamics is certainly a direct consequence of the first engineering function of the universe (as is Aristotle's dictum that nature abhors a vacuum) whereas the first two laws of thermodynamics (as well as the so called "zeroeth law") are direct consequences of the second engineering function. The first engineering function effectively dictates that we *have* a universe at all whilst the second engineering function (which could also be called *"the principle of parsimony"*)[18] ultimately determines the efficient *form* of the universe (its quantum format, the physical constants etc.) given the universe's prior obligation to exist.

---

[18] Not to be confused with William of Ockham's admirable razor, also sometimes called the "principle of parsimony". The second engineering function implies that although the universe (relatively speaking) is vast its very existence is a reluctant one.

These two engineering functions appear, superficially, to be at odds with each other (which is why the perfect symmetry of the radiation origin is spontaneously broken) but, as we have seen, because the second engineering function takes a *net* form (and should therefore technically be expressed as; *the net conservation of all symmetries*) this prima-facie contradiction is overcome. *Our universe is entirely the by product of this reconciliation of ostensibly opposed logical (i.e. apriori) requirements.*

Thus, although our universe has profound thermodynamic foundations (expressed as the three laws of thermodynamics and quantum physics), these foundations are *themselves* underpinned by still more fundamental and hitherto un-guessed logical principles, which are in fact logical derivatives of the former. This is why throughout this work I assert the critical importance of these two new postulates which are the key to our ultimate understanding of the purpose and form of the universe.

In conclusion then, the critique provided in this section helps to place the familiar cyclical hypothesis on far firmer foundations than ever before, as does the related analysis of curved time and associated symmetry conservations made in sections nine and ten. All of which are profound corollaries of the apriori character of net perfect order.

17. An Imaginary Cosmology.

The curved time model laid out in section nine has the effect of qualifying another major cosmological model (the only one we have not yet addressed) which is Stephen Hawking's version of Professor James Hartle's so called "no-boundary proposal".[19] According to this proposal the universe has no defined initial conditions (as is the case at the radiation origin) and as such no special rules apply that mark out one point in the space time continuum as being physically different (from a fundamental perspective) from any other. In other words, the same physical laws apply everywhere eternally.

This proposal is interpreted by Hawking in terms of a mathematical construct called "imaginary time" (imported because of its extensive use in quantum mechanics) to mean that space-time is finite and curved (in the imaginary dimension) and so has no boundary, in conformity with Hartle's suggestion. It thus has the unique distinction of being the only cosmology prior to my own which attempts to treat relativistic time (albeit only in its imaginary aspect) as a dimension of space. Although Hawking uses this overall construct probabilistically it amounts to little more than saying that space-time must be curved and without singularities, which is simply to assert what was already known. Nevertheless, since, as we have already seen, the universe *was* once a quantum scale entity, the application of wave-function analysis in the manner suggested by Hawking is technically feasible. But whether it is truly useful or not (since it merely reaffirms what we already know) is another issue.

The fundamental problem with the "imaginary" cosmology however stems from the fact that Hawking insists on treating the imaginary dimension as a physical reality, whereas the presence of the imaginary dimension in quantum calculations is more usually interpreted (following the work of Max Born) as indicating the presence of *probabilities* rather than literal quantities as Hawking seems to assume. Complex numbers are, after all, nothing more mysterious than ordered pairs of *real* numbers.

In the end the issue comes down to the correct interpretation of the wave-function in quantum mechanics. Hawking, influenced by Feynman's path-integral interpretation (itself somewhat in conflict with Born's more orthodox interpretation) defaults to a discredited literalistic interpretation, whereas a correct, that is to say *probabilistic* interpretation of the wave-function has the effect of undermining Hawking's interpretation of his own work. The correct interpretation of Hawking's imaginary cosmology is therefore dependent on the true interpretation of quantum mechanics itself, a subject we will animadvert to later and in greater depth (see section thirty seven et seq).

Hawking's cosmology with its literalistic interpretation of imaginary time is in any case rendered somewhat redundant by my own since what Hawking is able to do with his (I believe) misinterpretation of imaginary time is achieved by my own reinterpretation of relativistic "real" time, without the need to lend imaginary time an inappropriate physical significance. Hawking's perennial complaint that we are being narrow minded in not allowing a physical significance to imaginary time would hold more water were it not for the fact that the link between imaginary concepts and probabilistic analysis has been definitively established by the work of Born and others, thus rendering the need for a physical (i.e. literalistic) interpretation of imaginary time superfluous and, I suspect, unsustainable.

---

[19] J. B. Hartle and S. W. Hawking, Wave function of the universe. Phys.Rev, D:28 (1983) 2960.

18. The Hyper-Atom.

By treating time, on the authority of Einstein and, latterly, of Hawking as another dimension of space it becomes legitimate and even necessary to view the universe, like the energy that composes it, as being uncreated and undestroyed. It becomes, in a nutshell, an hyper-sphere. In fact, in a striking example of duality or a gestalt switch the universe can equally well be described as being in eternal recurrence (in effect a state of perpetual motion[20]) or, if time is correctly treated as a *spatial* dimension, as being an un-altering world sheet in the geometrical form of an hyper-sphere. These two descriptions are exactly equivalent.[21]

A further refinement of this latter concept leads to the postulation of the universe as an hyper-atom, its evolution through time uncannily resembling, in a fractal way, the structure of an atom. Under this description, the new state of matter identified at the twin quantum origins represents the hyper-nucleus, whereas each successive phase in the evolution of the universe – plasma, pulsar, galaxy and nova phases, as described by the hot big-bang model – represent successive hyper-rings or hyper-shells of the hyper-atom. In other words, each successive phase in the evolution of the universe through time, being less dense than the one preceding it, mirrors (in an informal way) the decreasing compaction in the outer electron shells of a microscopic atom.

The whole is literally an "a-tom" meaning "un-cut" in ancient Greek – i.e. an indivisible entity. Strictly speaking therefore and bizarre as it may sound, the universe is the *only* true atom. Attempts to analyze isolated aspects of the universe (notably through the practice of the inductive method) are what are partly responsible for the indeterminacy that is so precisely formulated by the uncertainty principle. This is because such isolations of the part from the whole are artificial in nature and so do violence to the fundamental unity of the hyper-spherical space-time continuum. It seems that the punishment for this inescapable perspective (which I identify as "universal relativity" and which is an inevitable corollary of cognition) is unavoidable indeterminacy and the consequent narcissistic illusion of free-will.

To our three dimensional eyes the hyper-nucleus appears to be a pair of quantum origins which, though symmetrical to each other are in reality mirror reversed, resembling, in effect, a super-particle pair. However, taking the hyper-dimensional perspective afforded to us by curvilinear time into account, the two quantum origins in reality form a *single* hyper-nucleus to our hyperatom. *In fact they represent the two poles of the hyperatom as a hyper-sphere.*[22] Thus it is possible to understand the quantum origins and the two phased universe as nothing more nor less than an hyper-sphere. The fact that the postulation of curved time *as well as* the postulation of an hyper-sphere entail twin nuclei (effectively equivalent to twin poles) seems to me a striking corroboration of *both* hypotheses (which, in reality are different aspects of the *same* construct), hypotheses which dovetail together perfectly in spite of their being highly counter-intuitive. The hyperatom is simply a final refinement of this overall model. Though it appears to

---

[20] There would be no external source of friction capable of preventing a closed universe, with net zero entropy, from exhibiting perpetual motion.

[21] The doctrine of eternal recurrence has of course been propounded many times through history, but it is only Einstein's theory which is capable of supplying the doctrine with much needed empirical foundations, (given that "eternal recurrence" and the "Block universe" are essentially identical models.)

[22] The mathematical properties of an hypershere are described by Riemann. The formula for an *n*-dimensional sphere is;

$$x_1^2 + x_2^2 + ... + x_n^2 = r^2$$

Given the implications of M-Theory it is reasonable to suppose that the universe is not a four dimensional sphere but an eleven dimensional sphere (or 10-sphere), with seven of these dimensions curled up to sub-microscopic scales.

Incidentally, the hypersurface area of an hypershere reaches a maximum at seven dimensions after which the hypersurface area declines towards zero as *n* rises to infinity.

be transcendental in nature (since it treats time as a spatial dimension) this model is clearly derived from the application of the inductive method.

19. The Rebirth of Rationalism?

It transpires therefore (in view of the "Hyper-atom" hypothesis) that the solution to Kant's apparently insoluble "cosmological antinomy" is that the universe is finite and boundless, like a sphere, a solution first suggested in the ancient world by Parmenides (and not even entertained by Kant) and in the modern world by Hawking. It is remarkable, I feel, and hitherto quite unappreciated, that the one possible rational synthesis of Kant's primary antinomy of pure reason is *precisely* the form adopted by the universe. It suggests that the universe is indeed rational, as the Rationalists suggested, but that its particular form can only be discovered empirically. In other words, the solution to an antinomy Kant considered to be irresolvable ironically ends up corroborating, in this instance at least, the primary thrust of his synthetic philosophy (synthetic, that is, of the two great systems of modern philosophy, Rationalism and classical Empiricism).[23]

Kant's error was to imagine that an antinomy, because it could not be solved by pure reason in isolation cannot therefore be solved at all. Although the correct hypothesis *could* be made by pure reason alone it could not ever be sure of itself. But by using reason *in concert* with the empirical method (itself rational) the antinomy can infact be resolved. This approach, wherein the empirical method is utilized as a special organon of reason I call neo-rationalism. It is different from Kant's synthesis of reason and empiricism (on equal terms) for two reasons.

Firstly it does not treat reason and empiricism as equal. Empiricism, though rational, is only a small *part* of Rationalist philosophy (the method part). Secondly, Kant's pessimism concerning the so called *limits of human reason* is severely qualified. This is evinced by the potential of a souped up species of rationalist philosophy (souped up, that is, by the empirical method) to solve the insoluble antinomies and numerous other epistemological problems besides.

Consequently, since we are able to see the relationship between Rationalism and Empiricism in much greater detail than Kant (who was the first modern to posit the interconnection) then I cannot count my position as quite Kantian either. Since I differ from all three systems in that I choose to interpret empirical method as a *special case* of rational philosophy (rather than coequal to it) it is therefore most accurate to describe my philosophical position as being *neo-Rationalist*.

The fact that the empirical method is *itself* rationally justifiable (since the problem of induction was solved by Popper) in turn implies that empiricism, correctly interpreted, is *not* a philosophy complete unto itself as the empiricists have always assumed, but is more correctly interpreted as a methodological corollary to (a correctly interpreted and fully modernized) Rationalism. This interpretation, if it is correct, amounts to the rebirth of rational philosophy itself and indicates that Kant's concerns about the limits of human reason (by which he meant reason unaided by observation) are only partially justified under the neo-rationalist schema laid out in this work. Therefore, as rational inhabitants of a rational universe we believe, contra Kant, that full gnosis is perfectly possible, albeit highly paradoxical in nature. It is the issue of what form a fully modernized Rationalism must take which constitutes the subject matter for the remainder of part one of this work.

---

[23] Kant's error, as I shall argue was to assume that Rationalism (which concentrates on apriori sources of knowledge) and Empiricism (which concentrates on aposteriori sources of knowledge) are therefore epistemologically equal. But this is merely an error of focus, since Kant is essentially correct in viewing both as necessary elements of a complete epistemology. In reality Empiricism is correctly seen as a *component* of Rationalism. They are not at odds with each other (as Kant saw) but neither are they co-equal.

20. Back to the Eleatics.

These speculations surrounding what I call the "hyperatom" perhaps inevitably remind us of what is the first block-universe model in history – that of Parmenides of Elea circa 500 BCE;

> "Only one story, one road, now
> is left: that it is. And on this there are signs
> in plenty that, being, it is unregenerated and indestructible,
> whole, of one kind and unwavering, and complete.
> Nor was it, nor will it be, since now it is, all together,
> one, continuous. For what generation will you seek for it?
> How, whence did it grow?
> … And unmoving in the limits of great chains it is beginningless
> and ceaseless, since generation and destruction
> have wandered far away, and true trust has thrust them off.
> The same and remaining in the same state, it lies by itself,
> And thus remains fixed there. For powerful necessity
> holds it enchained in a limit which hems it around,
> because it is right that what is should be not incomplete.
> … Hence all things are a name
> which mortals lay down and trust to be true –
> coming into being and perishing, being and not being,
> and changing place and altering bright colour."[24]

As elsewhere in pre-Socratic philosophy profound themes of modern physics are foreshadowed, including an intimation (almost a tenet amongst these philosophers) of energy conservation. However, what is particularly striking about Parmenides' poem from our point of view is that it envisages the block universe as a sphere;

> "And since there is a last limit, it is completed
> on all sides, like the bulk of a well-rounded ball,
> equal in every way from the middle."[25]

Of this vision Karl Popper has commented;

> "Parmenides' theory is simple. He finds it impossible to understand change or movement rationally, and concludes that there is really no change – or that change is only apparent. But before we indulge in feelings of superiority in the face of such a hopelessly unrealistic theory we should first realize that there is a serious problem here. If a thing X changes, then clearly it is no longer the same thing X. On the other hand, we cannot say that X changes without implying that X persists during the change; that it is the same thing X, at the beginning and at the end of the change. Thus it appears that we arrive at a contradiction, and that the idea of a thing that

---
[24] Parmenides, "The way of Truth" lines 8-42. Translated by Jonathan Barnes in "Early Greek Philosophy" P134-5. Penguin Press. 1987.
[25] Ibid. lines 43-5.

changes, and therefore the idea of change, is impossible.

All this sounds very abstract, and so it is. But it is a fact that the difficulty here indicated has never ceased to make itself felt in the development of physics. And a deterministic system such as the field theory of Einstein might be described as a four dimensional version of Parmenides' unchanging three dimensional universe. For in a sense no change occurs in Einstein's four dimensional block-universe. Everything is there just as it is, in its four dimensional *locus;* change becomes a kind of "apparent" change; it is "only" the observer who, as it were, glides along his world-line; that is, of his spatio-temporal surroundings…"[26]

It is perhaps striking that an empirical induction (courtesy of Einstein) should appear to confirm a deduction from logic, but this is not the only such example. It is also worth noting that the atomic hypothesis itself was first postulated by Democritus in the fifth century BCE on purely logico-deductive grounds, providing further evidence for the fact that inductive conjectures are ultimately the product not merely of empirical observation but of rational speculation as well. Consequently, the false cleavage between Rationalism and Empiricism that has riven modern philosophy and which did not exist for the ancient Greeks is, as Kant observes, unsustainable.

Popper also conjectures that Parmenides' model was motivated by the need to defend the idea of "being" against Heraclitus' doctrine of universal change, a doctrine which appears to lead to a sophisticated form of ontological nihilism;

"We step and do not step into the same rivers, we are and we are not."[27]

Popper additionally observes (what is widely recognized) that the Atomic hypothesis of Leucippus and Democritus was in turn motivated by the desire to defend the idea of motion against the theory of the Eleatics and in particular against the paradoxes of Zeno. It is, in consequence, the collapse of the classical atomic model (brought about by modern quantum mechanics) which has revived the possibility of a block-universe model *and* the Heraclitean theory of (what amounts to) universal relativity.

Indeed, these two visions are profoundly complementary, as this work hopefully demonstrates. The difference is one not of substance but of perception, a fact which would probably have disturbed Parmenides (a monist) rather more than it would have done Heraclitus[28]. Which vision one prefers

---

[26] Sir Karl Popper, "The Nature of Philosophical Problems and their Roots in Science." In "The British journal for the Philosophy of Science." 3, 1952.

[27] Heraclitus [B49a], translated Barnes loc cit, p117. According to Quine (*Word and Object,* p116) the paradox is solved linguistically by understanding that the concept "river" is held to apply to an entity across time quite as much as it does to one across space. This interpretation is however already implicit in Heraclitus' observation that we do indeed step into the same river more than once. Quine's linguistic solution is infact based on a misquotation of Heraclitus' which mistakenly has Heraclitus *forbidding* the possibility of stepping into the same river ever so many times. Furthermore, in what is a highly suggestive parallel Quine is led from his partial solution of the paradox to posit the possibility of a block-universe as the logical corollary to his own solution (ibid, p171).

Of this we may say two things. Firstly, a supposedly *linguistic* solution of a classical philosophical problem (a pseudo-problem perhaps?) has significant consequences for physical theory – to wit the positing of a block-universe, (in which case, presumably, the problem is not merely linguistic after all?) From this we may deduce the important principle that since the problems of physics and language elide together it is therefore wrong to posit a determinate border between philosophy and the sciences as linguistic philosophers implicitly do. The fundamental error of linguistic philosophy whose unsustainability is illustrated by just this one example, is one of over compartmentalization.

Secondly, in the historical debate as to which came first, Heraclitus' system or Parmenides' it seems apparent, from Quine's scholastically naïve reaction, that those of us who suspect that Parmenides' system arose as a direct response to this paradox may well be correct after all, (in much the same way that we believe Atomism to have arisen in response to a logical inconsistency in Eleatic monism). I also happen to suspect that the ultimate source of Zeno's paradoxes lies in a generalization of Heraclitus' paradox of the river, but a fuller treatment of this must wait until section twenty two.

[28] It was Heraclitus after all who said; "Changing, it rests." [B84a] Ibid. This suggests either that he was already familiar with Parmenides' theory or else, (more likely,) that he deduced this paradoxical situation as a logical consequence of his own doctrine of perpetual flux. Parmenides, we suspect (rather like Quine) merely unpacked one aspect of what Heraclitus was

depends solely on whether one elects to treat time as a dimension in its own right (as one should if one aspires to completeness) or not. If one does then one arrives at a block-universe model. If one does not then Heraclitus' vision (in which objects are transmuted into processes and everything only exists relative to everything else) becomes inevitable, serving, in effect, as a *limiting case* (in three dimensions) of the four dimensional vision of Parmenides and Einstein;

> "The world, the same for all, neither any god nor any man made; but it was always and will be, fire ever living, kindling in measures and being extinguished in measures."[29]

But even this vision (of an eternal universe composed entirely of "fire") is strikingly akin to that arrived at empirically by Einstein. For example, it suggested to Heraclitus a doctrine which is uncannily similar to that of mass-energy inter-convertibility;

> "All things are an exchange for fire and fire for all things, as goods are for gold and gold for goods."[30]

Although the ideas quoted in this section *were* inductively arrived at (through a combination of observation and logical reasoning) and although they are largely correct we may nevertheless speculate that it was because of the absence of an inductive procedure incorporating the concept of the *testing* of hypotheses ("putting nature to the question" to quote Francis Bacon) that these ideas proved incapable of maintaining the centrality they deserved. This is not of course the fault of the ancient "phusikoi" themselves. Given the absence of modern technology ideas such as theirs would seem to be metaphysical whereas in reality they *are* susceptible to empirical testing, further evidence that the divide between the metaphysical and the empirical is not as clearly defined (except in retrospect) as naïve empiricists and inductivists sometimes like to suppose. Nevertheless it was not until the incorporation of testing procedures (initiated by Archimedes at the very end of the Hellenistic era) that modern Inductivism (as a method) became (in principle) complete.

---

suggesting. Indeed, Quine's stated holism seems to have originated from a problem situation startlingly similar to that faced by Parmenides, whose solution was *also* holistic in nature (i.e. *monism*). And this coincidence is further circumstantial evidence for our contention (see the following note) concerning the conceptual similarity of Heraclitus' great system to modern quantum physics (and also concerning his historical priority viz Parmenides).

[29] Heraclitus [B30] Ibid p 122. It is clear that Parmenides' overall solution solves the paradox of change identified by Popper and in a way which anticipates Quine. Nevertheless this fragment of Heraclitus, echoing B84a, effectively iterates Parmenides' vision of eternal stasis, inspite of the fact that Heraclitus is usually interpreted as the first "process" philosopher. Perhaps it would be more accurate to regard the rival Eleatic and Atomic systems as *decompositional elements implicit within Heraclitus' system*, which, of all pre-modern systems, is closest in nature to modern quantum physics. And yet each of these three proto-empirical systems were constructed rationally, in response to primarily *conceptual* paradoxes.

[30] Heraclitus [B90] Ibid p123. Although the great contemporaries Thales and Anaximander must be considered the primary originators of Greek philosophy (that we know about) it was, I believe, Heraclitus who determined the dominant *issues* for the great period of philosophy, the greatest in history, which culminates in the towering presences of Plato and Aristotle and their associated schools. Heraclitus, as it were, upsets the very *stability* of the Greek mind, but in a deeply creative and consequential fashion.

21. The Rational Basis of Inductive Philosophy.

As mentioned in the previous section Democritus is reputed to have postulated the atomic hypothesis for purely *logical* reasons in order to obviate the problem of infinities.[31] For this interpretation we have no less an authority than that of Aristotle himself;

> "Democritus seems to have been persuaded by appropriate and scientific arguments. What I mean will be clear as we proceed.
>
> There is a difficulty if one supposes that there is a body or magnitude which is divisible Everywhere and that this division is possible. For what will there be that escapes the Division?
>
> … Now since the body is everywhere divisible, suppose it to have been divided. What will be left? A magnitude? That is not possible; for then there will be something that has not been divided, but we supposed it divisible everywhere…
>
> Similarly, if it is made of points it will not be a quantity… But it is absurd to think that a Magnitude exists of what are not magnitudes.
>
> Again, where will these points be, and are they motionless or moving?
>
> … So if it is possible for magnitudes to consist of contacts or points, necessarily there are indivisible bodies *and magnitudes*."[32]

Leaving aside the embarrassing fact that this still remains a perfectly valid criticism of the point particle "Standard Model" of modern physics (though not of string theory) this passage makes it absolutely plain that purely *logical* considerations generated the atomic hypothesis (the case is otherwise for modern quantum theory however).

There is thus *nothing* empirical about the original conjecture of atomism and yet it is still clearly an *inductive* hypothesis, ultimately susceptible to testing and observation. This suggests to me that though inductive conjectures may, in extreme cases, be triggered purely by an observation or (conversely) purely by logical reasoning (as in the case of the atomic hypothesis) *most* inductive conjectures are (as in the case of the quantum hypothesis) the product of a synthesis (often a complex one) of *both* sources.

The historical importance of the example of Democritus (and of Parmenides for that matter) is that it reveals the insufficiency of making observation *alone* the principle source of inductive conjectures, an error (usually, though perhaps unfairly, associated with Francis Bacon) of primitive or naïve inductivism, of inductivism in its early conceptual stages.

---

[31] By an odd coincidence (or perhaps not so odd given the considerations of this section) it was for precisely the same reason (to avoid the classical prediction of infinities) that Max Planck postulated the quanta in 1900, thus effectively reestablishing the atomic hypothesis on much the same grounds adduced by Democritus.

[32] Aristotle, "*On Generation and Corruption*" 316a 13-b16. Translated Barnes, loc cit, p251. (Emphasis added).

22. Zeno and Quantum Physics.

In essence therefore the atomic hypothesis appears to have been postulated as a basically correct (if incomplete) riposte to the paradoxes of Zeno. The full and correct response is, in principle, supplied by quantum physics (including relativity theory) and further suggests that the "engineering function" of quantization is indeed the avoidance of paradoxes associated with motion. In other words, what I call the "quantum format" of the universe is not merely an exotic and fundamentally inexplicable cosmological detail, but a logically necessary feature of physical reality, a fact first intuited by Leucippus and Democritus (we assume) two thousand five hundred years ago.

The paradoxes of Zeno are not, after all, paradoxes of mathematics (except incidentally), but rather of time and motion and hence of physics. Since classical physics was based on the same epistemological paradigm as classical mathematics it proved incapable of completely accounting for them. The fact that quantum physics *does* provide the remaining elements of a correct response is therefore a further indication of its fundamental (as distinct from heuristic) nature.

The error of interpreting time and space as purely *classical* phenomena (which is also the error implicit in the concept of the infinitesimal) was spotted by Aristotle more than two thousand years ago;

> "It depends on the assumption that time is composed of instants; for if that is not granted the inference will not go through."[33]

The calculus, in other words, is a classical tool, but time and space are non-classical. The calculus assumes spatio-temporal continuity, but time and space are *discontinuous* at a fundamental level. A line is *not* infinitely divisible, there is no "instantaneous rate of change", since this phrase encapsulates a contradiction in terms. Hence there is no such thing as an *infinitesimal* in nature. Instead, nature merely *tends towards* behaving as if there were an infinitesimal *as the limit is approached* – a limit defined in nature by the Planck time and the Planck length - which therefore represent the true limit of the divisibility of physical lines.

Thus the concept of infinity is as empty of *physical* significance as is that of an infinitesimal. Though mathematically significant the idea of a convergent series has no physical significance and so solving the paradoxes by graphing space with respect to time ignores the fact that the paradoxes apply (in principle) to motion through *time* quite as much as they do to motion through space.[34]

Consequently, Einstein's block universe (corresponding to Parmenides system) and Planck's quanta (corresponding to Democritus' solution) far from being *rival* solutions, as was assumed to be the case in classical philosophy, turn out to be *complementary* to one another. The fact that nature appears to opt for *both* of them suggests that the correct modern solution to the ancient dispute of the Eleatics and the Atomists (rather like Kant's solution to the dispute between the Rationalists and the Empiricists) is a *synthetic* one, something not suspected before.

It follows therefore that whilst the universe is indeed changeless (a la Einstein and Parmenides) the subjective *experience* of change is facilitated by the quantization of time and space. However, it is important to notice that although quantum theory restores conditions allowing for the possibility of the *subjective* experience of motion through time and space, it is the *apriori* principle of symmetry (i.e. the

---

[33] Aristotle, *Physics*, 239b5-240a18. Barnes op cit, p156.

[34] The most famous example of a convergent series is the following infinite series which converges to one;

$$\sum_{n=1}^{\infty} \frac{1}{2^n} = \frac{1}{2} + \frac{1}{4} + \frac{1}{8} + \frac{1}{16} + \ldots = 1$$

Though mathematically significant the infinite series has no *physical* significance, since space and time are not infinitely divisible because, in an *absolute* sense, they do not exist.

very parsimony of the universe, its reluctance to allow substantial existence) which translates potential chaos (unrestricted quantum indeterminacy) into ultimate net *order* – thereby allowing for the manifestation of *causality* (and hence the *subjective* experience of orderly *continuity*) at the *classical* scale.

Without the correspondence between classical and quantum scales supplied by the phenomenon of *symmetry*, orderly experience as we understand it would not be possible at all.[35] Thus it is no great surprise that the solution to the mystery should involve the aforementioned synthesis. Indeed, in retrospect it seems to require it.

These problems and paradoxes surrounding motion can therefore be interpreted as pointing to a fundamental disjuncture between the classical or ideal world of (classical) mathematics (and classical physics) and the discontinuous and relativistic or *indeterminate* reality that is the case. As such the correct and final form of their explication was not *comprehensively* available to us (at least in principle) prior to 1900 (or rather 1915).

Democritus solves the problems of motion vis-à-vis space but it is modern quantum physics (including relativity theory) which completes the solution and confirms its validity, extending it to include time as well.

---

[35] Without the breaking of primal symmetry the experience of motion (and thus of existence) would not be possible. This accords with Curie's aforementioned principle that phenomena are dependent on the breaking of symmetries for their existence. Fortunately, due to the apriori necessity of the *first* engineering function of the universe symmetry breaking is deducible, *apriori*, as well as being observable aposteriori. It is the second apriori engineering function we must thank however for keeping things *orderly*.

23. Why Zeno's Paradoxes are Both Veridical and Falsidical.

The paradoxes (along with the block universe of Parmenides) therefore supply us with an early inkling of the purely *relative* nature of space and time and that space and time are not absolute (Newtonian) entities.

In undermining the concept of absolute motion the paradoxes instead point to universal relativity, the block universe and ultimately to non-locality itself. They therefore disconfirm Parmenides' monistic ontology whilst at the same time confirming his physics.

Furthermore, in as much as Parmenides' block universe is implicit in Heraclitus' doctrine ("changing it rests") it is equally fair to consider the paradoxes as a corollary of Heraclitus' system as well. Indeed, their historical context and intellectual content imply them to be a generalization of the paradox of the river stated earlier.

Thus, in essence, all of Greek philosophy in the subsequent period, including Parmenides, Zeno, Democritus, Gorgias, Protagoras, Plato, Aristotle and even Pyrrho already lies, densely packed, like a coiled spring in these uniquely great aphorisms of Heraclitus. The great phase of Greek philosophy is in large part simply the full uncoiling of this spring.

In effect we are faced with what might be called a *meta-paradoxical* situation in which quantization appears to *refute* the paradoxes, but relativity *affirms* them. The peculiar difficulty in resolving them stems, I believe, from this *inherently* indeterminate status, a situation which reminds us somewhat of the inherently indeterminate status of the continuum hypothesis in mathematics.

Since there is no such thing as *absolute* space and time the paradoxes are veridical, but in order to "save the phenomenon", i.e. in order to allow there to even *be* phenomena at all quantization and relative space and time are apriori requirements. And thus the paradoxes are falsidical as well.

This perplexing situation becomes a good deal less perplexing when we recognize it as, in effect, a corollary of the situation described earlier concerning the tension that exists between what I have identified as the two engineering functions of the universe. In essence the *indeterminate* status of the paradoxes is an inevitable consequence of this tension, a tension between being and nothingness.

There cannot (by definition) *be* absolute nothingness, ergo there must be *phenomena* of some sort. And the consequence of there being phenomena is the relativity of space time and quantization. These are pre-requisites of phenomena and phenomena (per-se) are the *apriori* consequence of the two (apriori) engineering functions of the universe. Thus the simultaneous veracity and the falsity of the paradoxes is unavoidable. As Heraclitus puts it; "We are and we are not."

And so, in conclusion, the paradoxes demonstrate that there can be no such thing as *absolute* space and time, but they do *not* prove that *relative* motion is impossible. Hence they are both veridical *and* falsidical.

24. The Super-Universe.

A feature of the radiation origin that we have down-played hitherto is that it represents a source of unlimited potential energy. This fact in turn opens up the possibility of a plurality of universes with similar properties to our own. An ensemble of such universes or hyper-spheres might reasonably be termed a "super-universe"[36]. This hypothesis is similar, but significantly different, to Everett's popular "many universes" hypothesis, which stems from a naive literalistic interpretation of the Schrödinger wave-function and which seems to me to lead to the inevitable and conceptually flawed assumption of infinite energies. Therefore and given that the literal interpretation of the wave-function was discredited in the 1930s by Max Born in favour of the statistical interpretation I feel that it is unavoidably necessary to discard Everett's "many-universes" hypothesis as invalid. This is because Born's interpretation and the many universes hypothesis cannot logically coexist. Likewise the "Schrödinger's cat" paradox and the hot (yet sterile) issue of wave-function collapse are rendered irrelevant by a correct understanding of Born's statistical interpretation of the wave-function.

The super-universe hypothesis is not so easily discarded however. The super-universe would presumably exist in a "steady-state" of net zero vacuum energy (much as our own universe seems to do), but, due to vacuum fluctuations of the sort that created our universe, countless *other* universes could in theory *also* be created, out of the perfect symmetry of absolute nothingness (analogous to an infinite number of dots existing on a sheet of infinitely sized paper). This hypothesis would also provide a simple solution to the problem of tuning, since, if so many diverse universes exist it would then hardly be surprising that at least one of them should exhibit the awesome fine tuning that is apparently required to explain the existence of conscious life, (see section thirty six for a further discussion of this issue).

It is important to admit therefore that the super-universe model is *not* ruled out by quantum theory, which is our guiding light in these matters (and presumably always will be). Nevertheless it seems to me that this enticing model *is* probably ruled out by deeper logical principles of the sort that we have already discussed in this work, principles which trump even quantum theory. Most notably, I suspect that the super-universe would very likely (though not certainly) violate what I have called the first engineering function of the universe. This is because it opens up the possibility of an infinite number of universes and yet it is precisely the problem of infinite energy that caused us to postulate this engineering function in the first instance. Although it could still be argued that the net energy of the super-universe remains zero, it is not obvious to me why, under this model, there would not be an infinite number of universes at any given time – a solution which is ipso-facto absurd (like the idea of infinitesimals) and which therefore militates strongly against the super-universe hypothesis.

Contemporary cosmology, as our above analysis shows, suffers, like quantum field theory, from the hidden presence of lurking infinities. The difference is that quantum field theory (with its ad hoc dependence on methods of "renormalization") is at least aware that this situation constitutes a fundamental contradiction of their system, but cosmologists continue to make use of concepts that raise the same problem of infinity (multi-verses, singularities, worm-holes etc.) without apparently being troubled by the ontological implications of this. They too commonly use these terms as though what they referred to were given, empirical facts rather than indicators of a fundamental theoretical problem. As a result we see the descent of much contemporary cosmology into uncritical and self indulgent metaphysics. The chief advantage of Hyper-symmetric cosmology however is the effortless way it closes out most if not all of these problems of infinity whilst at the same time retaining a close connection with the underlying theoretical constructs; *General Relativity* and *Quantum Theory*.

---

[36] British astronomer Martin Rees has also coined the term "multiverse" for much the same concept.

25. Is the Universe in a Spin?

Just as Einstein's equations can be solved for an expanding universe, they can also be solved for a rotating one as well. Gödel's rotational solution, discovered more than half a century ago, was subsequently demonstrated to be inapplicable to our particular universe, but the possibility of similar rotational solutions existing which are not so inapplicable cannot be ruled out. Part of the problem with the hypothesis has been the fact that no one has been able to suggest what the universe might be rotating relative to. As a result of these two difficulties the hypothesis has been relegated to an undeserved obscurity almost since its inception. Nevertheless, it is apparent to cosmologists that all very large objects in the universe do in fact exhibit spin, just as do all fundamental particles. Were it not for the aforementioned problems it would seem entirely logical to attribute this property to the universe itself.

Apart from Gödel's solution two other arguments exist in favour of the rotational hypothesis that have, I believe, never been considered before. The first argument pertains to the quantum origin which, as our analysis in section two demonstrated is a spinning particle (a fermion) at 10^{-43} seconds. This spin could well form the basis of subsequent rotation and suggests that the possible angular momentum of the universe is one of the many by products of spontaneous symmetry breaking at the radiation origin.

The second argument in favour of a rotating universe pertains, appropriately enough, to the end or fate of the universe. In our model, as has been argued, the universe is closed. This model assumes an eventual gravitational collapse of the entire universe back to the state of the quantum origin. This in turn implies that the universe at some future date will become a black-hole. Since *all* black-holes possess angular momentum it therefore follows that the universe when it enters this state *will* possess angular momentum as well. And given that we have established a spin associated with the quantum origin we would expect this angular momentum of the universe to be non-zero. Though this is not proof that even a closed universe is necessarily spinning in its current state, it at least raises the possibility of this. All of which seems to confirm our earlier intuition that an object as large as the universe is likely to have angular momentum.

Pressing the black-hole analogy a little further it would seem that, like a black-hole, any closed universe must have an electrical charge associated with it as well. It is possible that this charge relates to the observed matter/anti-matter imbalance in the universe (the C in C.P.T), but this is almost certainly not the case. Perhaps a better solution is to treat the universe in its highly dense contracting phase as an electrically neutral black-hole. The only alternative to this (to my mind highly satisfactory) interpretation requires a revision of the quantum origin model to give the quantum origin electrical charge as well as spin, but this does not seem to be as satisfactory a solution on a number of counts.

Another major problem for the rotational hypothesis concerns what the universe may be spinning relative to given that it cannot spin relative to itself. Bearing in mind our analysis in the preceding section it would seem rational to assume that the universe is spinning relative to *other* universes in a super-universe ensemble. But this startling conclusion is unsatisfactory since it resurrects the spectre of infinity and the attendant violation of the first engineering function discussed in the last section.

A more economical solution to this problem that does not violate any fundamental logical principles involves the interpretation of spin as representing additional evidence in favour of the two phased universe or "hyperatom" model discussed earlier. According to this interpretation the angular momentum of the universe is occurring relative to that of the *second* phase of the universe (and vice-versa), leading to a net conservation of angular momentum in conformity with the second engineering function of the universe. Although this is a bizarre solution, it has the advantage of being economical and solves the various problems associated with the rotational hypothesis whilst fitting in seamlessly with our overall cosmological analysis. Furthermore it avoids the problems of infinity raised by the only alternative (the super-universe solution) to this possible phenomenon. Accordingly we may legitimately assume that there is just a single hyper-spherical universe and no fundamental logical principles are violated.

Spin also provides a final tweak to our overall model of the universe as an hyperatom since it suggests that the universe is not strictly speaking an hyper-sphere (as hitherto assumed for the sake of simplicity) but is actually elliptical in form since spin normally leads to elliptical effects. The hyperatom is thus not an hyper-sphere but an *hyper-ellipsoid*, given the reasonable assumption of spin. This spin furthermore is hyper-spatial in nature.

26. Black-holes Unmasked.

Our treatment so far leaves only one well known cosmological anomaly unaddressed, which is that posed by dark stars or so called "black-holes". According to the current relativistic paradigm, stars above two solar masses undergoing gravitational collapse (having spent their hydrogen fuel) inevitably collapse towards a dimensionless point of infinite density, i.e. a singularity. However, this paradigm is undermined by a couple of key points, the most glaring of which is that the rules of quantum theory forbid such extremal collapses, since they transgress the fundamental limits imposed by the Planck dimensions. From this fact we may deduce that although General Relativity is useful for indicating the *existence* of black-holes to us, it is nevertheless incapable of illuminating the sub-atomic structure of the phenomena whose existence it flags up. This situation is similar to the case of the radiation origin discussed earlier.

The other problem with the singularity model pertains to the fact that all black-holes have *finite* "event horizons" (i.e. circumferences) associated with them whose radius is (in *all* cases) precisely determined by the *mass* of the black-hole. Of this we might facetiously say that it points to an anomaly in our anomaly since an infinite singularity should not give rise to a finite, mass dependent, circumference in this way.

These facts clearly tell us that whatever is going on inside a black-hole *cannot* be a singularity. They also point to the need to turn to quantum theory in order to discover what is really going on. Specifically, what we are seeking is a culprit for the phenomenal mass-density of black-holes which in *all* cases manifest the same fixed and finite density of around a billion metric tons per 10^{-13} cm^3 (which is approximately the volume of a neutron).

Of greatest significance in this respect must be the class of particles known as "quarks" since, apart from the relatively lightweight class of particles known as "leptons", they form the entire bulk of all significant visible mass in the universe. We would think therefore that an analysis of the properties of quarks would be of the utmost importance in the context of understanding the component basis of black-holes. And yet such an analysis nowhere exists, which is why we are left with the palpably absurd "singularity" paradigm of a black-hole interior. Yet the question abides; why, if matter does collapse uncontrollably towards a singular point is the average mass-density of a black-hole of any given size always fixed? And why, indeed, do black-holes possess any finite size at all? Let alone one strictly correlated to their mass?

I suspect, therefore, that the key to solving the mystery of a black-hole's mass lies with string-theory, itself perhaps the most sophisticated and coherent expression of the quantum theory to date. Under the traditional paradigm of particle physics quarks, like all other particles, represent dimensionless points and so are susceptible to the kind of unlimited compression implied by General Relativity theory. But in string theory, which respects the fundamental character of the Planck dimensions, a limit to this compression is provided by the Planck length at 10^{-33} c.m.. A quark, in other words, cannot be compressed below this size, a fact amplified by the existence of the Pauli Exclusion Principle, which should operate towards counteracting any such infinite compression. The finite compressibility of quarks thus provides a natural limit to the collapse of a black-hole which, notwithstanding the Hawking-Penrose singularity theorems, simply cannot collapse to a singularity. These purely mathematical theorems, being products of the internal analysis of the General Relativity formalism, are ipso-facto superceded by an analysis from the perspective of quantum theory of the sort suggested here. By contrast, a proper application of string theory has the twin effect of ruling out the possibility of singularities (including so called "wormholes") and of supporting the alternative "quark" hypothesis offered here.

A further implication of a string analysis is that quarks *can* compress sufficiently to allow *stable* densities of the order of a billion metric tons per 10^{-13} cm^3. In fact, given this uniform density, it should be possible to calculate just how many quarks comprise the interior of black-holes and also the precise degree of compression they experience. This latter seems a particularly significant data point given the uniform nature of a black-hole's average density. At any rate, compressed (at least potentially) to the dimensions of a string, far more quarks can be expected to inhabit the typical area of a neutron (10^{-13} cm^3), than is *normally* the case. Thus the mass density of a black-hole can be expected to be far greater than that of a neutron star (whilst nevertheless remaining finite) which is indeed what we observe.

Furthermore, a radical hypothesis can be conjectured to the effect that the uniform density of black-holes is to be associated with quarks in their unconfined state. Under normal conditions quarks always appear in doublets or triplets, a phenomenon known as quark confinement, but at sufficiently high energies or densities (as, presumably, in a black-hole) quark deconfinement may obtain, hence the aptness of the name "quark star".

Further corroboration for this hypothesis (that black-holes are composed of highly compacted or possibly unconfined quarks) is perhaps provided by the study of the birth of hadrons in the very early universe (baryogenesis). Under the favoured G.U.T. scale theory, enormous numbers of quarks (hundreds of times more than presently exist) are believed to have sprung into existence as a by-product of G.U.T. or strong-force symmetry breaking at around 10^{-29} seconds after the beginning. The density of the universe at about this time would have been extraordinarily high, akin to that of a black-hole and yet there can be no doubt that the mass-density was composed of quarks.

In any event what I call the "Quark Star" hypothesis has the effect of restoring rationality to the study of black-holes since all infinities and singularities (including the pernicious "worm-hole" singularity) are eliminated by it. It also restores a degree of familiarity since it becomes apparent that what goes on within the sphere of a black-hole can in principle be described within the admittedly imperfectly understood frameworks of quantum chromo-dynamics and string theory. As such and given Professor Hawking's discoveries concerning black-hole radiation perhaps we should consider renaming these phenomena quark stars?

This new paradigm possesses the additional advantage of allowing us to identify a clear continuity between neutron stars (collapsed stars with a mass above the Chandrasekhar limit but below two solar-masses) and black-holes. Under this paradigm the Schwarzschild radius possesses no special ontological significance except to demarcate the point at which a neutron star (already very dark) ceases emitting radiation entirely and so officially becomes a black-hole or quark star. Apart from this the two phenomena remain much the same (unless the quark deconfinement model is correct) in that both are composed entirely of quark triplets with a ratio of two "down" quarks for each one "up".

Indeed, it is surely foolish to believe, as we have been obliged to under the old paradigm, that a collapsing star above two solar masses goes from being a perfectly rational, albeit extra-ordinarily dense, neutron star one moment into a mind-boggling and irrational "singularity" the next. This is clearly ludicrous and merely indicates our lack of understanding as to what the Schwarzschild radius really implies, which is, in actual fact, remarkably little.

27. Baryothanatos.

Thus it is impossible for a black-hole to acquire so much mass that it undergoes a phase transition into a singularity. This is primarily because the acquisition of ever more mass translates into an ever larger Schwarzschild radius. This in turn implies a uniform processing and distribution of quarks within the event horizon of a black-hole. If the hadron postulation were not correct then why would a black-hole's event horizon expand uniformly with its mass? And what culprit for this mass could there be other than quarks? Having disposed of singularities there simply is no alternative to this hypothesis.

It is important to point out at this juncture that "black-hole" (or indeed "quark star") is in reality a *generic* term and as such describes not just collapsed stars but also quasars, active galactic nuclei and sundry other types of object. Nevertheless, excepting variations in size, electrical charge and angular momentum (i.e. the properties of their constituent particles – *quarks*) all black-holes are quintessentially identical. A black-hole is thus, in a sense, a giant *hadron* and as such is incredibly stable and predictable. This is because (as Hawking and Bekenstein have demonstrated) quark stars represent the closest approach to thermal equilibrium allowed for by the laws of physics.

Notwithstanding the foregoing remarks there is one case of a black-hole which is capable of undergoing transition to a new state of matter. But this is a unique instance (that of a collapsing universe) and can only occur in the context of a closed universe. *In no other case* (as our analysis of baryogenesis showed) is mass density sufficient to cause a phase transition in a quark star. And this is an important point to stress, given the misunderstandings inherent in the orthodox interpretation of the Hawking-Penrose singularity theorems.

What then are the characteristics of this unique case? As the universe collapses it may be expected to generate a series of black-holes which subsequently merge to form a super-massive black-hole, in conformity with Hawking's area increase law. This super-massive black-hole will, in spite of its size, continue to obey Schwarzschild's equation (relating its radius to its mass) as, with increasing velocity, it swallows the entire mass of the universe.

Only in the last millisecond, when the process of baryogenesis is effectively reversed, will pressure densities annihilate even hadrons (a process we might justifiably term *baryothanatos*) triggering a phase transition. But this transition will not be into a singularity (as the Hawking-Penrose singularity theorems might lead us to suppose) but rather into the highly symmetrical *new* state of matter identified as the *quantum origin*. This phase of matter, incidentally, oscillating at 10^{28} eV (i.e. the Planck energy) can *also* be characterized as the highest possible string vibration of all.

At this point, because of the annihilation of quarks (baryothanatos) the universe will *cease* to be a black-hole and, having contracted to its minimal dimensions (of length, time and density) newly converted energy (converted by the process of baryothanatos) will then exist to power a renewed cycle of expansion (the "bounce" necessitated by quantum theory) as previously suggested. This is because Baryothanatos effectively removes the gravitational pressures that had previously resisted a renewed cycle of expansion. Thus baryothanatos is crucial to the functioning of the cyclical model offered in this work.

Consequently, the pressures that are capable of overcoming the baryonic structure of a black-hole are only free to exist circa the last 10^{-29} seconds of a contracting universe and do not exist in any other circumstances. Which is why, in all cases (according to quantum theory), a black-hole is incapable of collapsing to a singularity.

28. The Conservation of Information.

> "I believe that if one takes Einstein's General Relativity seriously, one must allow for the possibility that space-time ties itself up in knots – that information gets lost in the fold. Determining whether or not information actually does get lost is one of the major questions in theoretical physics today."[37]

The solution to the problem raised by Hawking in the foregoing passage lies in recognizing the deep connection that exists between thermodynamics and information theory, a connection that was first indicated by Claude Shannon in the form of his well known equation linking information with entropy. It is the implication of Shannon's law that information and entropy expand and contract in lock step with each other, which in turn implies that information must be conserved, just as energy is in the first law of thermodynamics. Thus the solution to Hawking's conundrum lies in the postulation of a law of information conservation to complement Shannon's law of information entropy. This might reasonably be called the *second* law of information theory, forming, alongside Shannon's law, a perfect mirror to the two principle laws of thermodynamics. Indeed, these two laws of information theory should rightfully be seen as special cases of the principle laws of thermodynamics on which they depend. But what does a law of information conservation mean in practice and how does it help us to resolve the conundrum posed by Stephen Hawking?

In the context of Hawking's conundrum the law of information conservation implies that any information gratuitously destroyed in a black-hole *balances out* information that is equally gratuitously *created* elsewhere in the universe – notably out of the "white-hole" constituted by the radiation origin, but also out of lesser "white-hole" phenomena such as stars and supernova explosions. From the broader perspective afforded to us by the second law of information theory we are thus able to see that the cosmic accounts remain balanced and Hawking's conundrum vanishes without the need to hypothesize, as he and Penrose do, that information is somehow conserved within an event horizon, destined to represent itself through some unknown mechanism at a later date. From what we now know about so called "black-holes" (i.e. quark stars) this is clearly nonsense as well as unnecessary. Indeed to say, as Roger Penrose does, that information is "lost" in a singularity[38], is equivalent to saying that "energy is lost" which contravenes laws even more fundamental than those of Einstein's General Relativity on which he and Hawking consistently over-rely.

A broader implication of Shannon's under-appreciated interpretation of the Boltzmann equation is that where entropy equals net zero (as I believe is the case in a closed universe) information must *also* balance out at net zero as well. Shannon's interpretation therefore implicitly defines existence itself as the continual processing of information such that the maximal point of expansion of the universe – being the high-water mark of positive entropy – is ipso-facto the high-water mark of information as well.

It also explains why quantum mechanics in its original form (Heisenberg's matrix mechanics) expresses itself in a binary format and D.N.A. takes a fully digitized quaternary format. This is because both fundamental phenomena (which define the binary code of the universe and the basic digital program for life itself) are ultimately only information.

---

[37] Stephen Hawking - Lecture at the Amsterdam Symposium on Gravity, Black-holes and Strings. 1997.
[38] "The Nature of Space and Time". S. Hawking and R. Penrose. Princeton U.P. 1996.

Shannon's law, furthermore, enables us to identify precisely *what* is undergoing disorder in Boltzmann's statistical reinterpretation of the second law of thermodynamics. The correct explanation is that *information* is what is being disordered. Shannon's law can thus correctly be seen as continuing the process of the reinterpretation of mass-energy *as* information (in effect the transition from an *analogue* to a *digital* interpretation of physics itself), a process begun by Boltzmann and hopefully nearing a conclusion at the present time.

The existential implications of this general reinterpretation of physics are also surely enormous given that information is simply the objective form of what we subjectively call "meaning". Thus if all information is ultimately conserved in the universe at net zero it therefore follows that the net *meaning* of the universe as a whole must *also* be net zero as well, irrespective of our subjective prejudices to the contrary. Consequently, and in antagonism to all our cultural assumptions hitherto, empiricism *does* have a profound impact on what can be said concerning subjective matters of existential meaning and purpose. Indeed, in many ways, it seems to have the final say.

29. The Third Law of Information Theory.

Indeed our earlier postulate, the first engineering function of the universe (dictating the unavailability of zero or infinite energy) also becomes clearer when expressed in terms of information theory (which is in effect the digital remastering of physics). Under this transformation the first engineering function simply asserts that zero (or infinite) *information* is unavailable to the universe.

Why is this an illuminating restatement of the first engineering function? Because it makes the logical necessity of the engineering function crystal clear. To put the matter bluntly; if there were no information in the universe (in effect, nothing in existence) then the *absence* of information would *itself* be an item of information. Ergo, there cannot *be* zero information in the universe. And this is *exactly equivalent* to saying that there cannot be zero energy in the universe. This is because zero energy implies zero entropy, which latter is forbidden by the third law of thermodynamics as well as by quantum mechanics (which can be defined as the mechanism by which the laws of thermodynamics express themselves). To put it still another way; the universe must be eternal since there can never have been a time when vacuum energy was not. Thus quantum mechanics and hence the laws of thermodynamics are perpetual.

In some respects it is possible to see the first engineering function as a simple restatement of the third law of thermodynamics or Nernst's heat theorem (which postulates that zero *entropy* is unattainable because, paradoxically, it entails infinite energy). To be precise, the first engineering function (which is a statement about *energy)* is a *derivative* of the third law of thermodynamics (which is a statement about *entropy*). The derivation is legitimate because zero entropy (aside form entailing *infinite* energy) also implies zero movement and hence zero heat, i.e. zero energy. To put it differently, it takes infinite energy to remove *all* energy from a system, ergo the postulation of the first engineering function of the universe. This postulation, furthermore, has a far wider import than the rarely referenced third law of thermodynamics.

By a similar argument the *second* engineering function of the universe (the conservation of all symmetries) can be interpreted as a disguised derivative of the *first* law of thermodynamics. Nevertheless the important point to remember is that the laws of thermodynamics are ultimately dictated by the two engineering functions of the universe, rather than vice-versa as might be supposed. This is because thermodynamics and its epigone, quantum physics, are *consequences* of an underlying logical necessity which is in effect defined by the putative engineering functions. The engineering functions, in other words, are *generalizations* of the (already highly general) laws of thermodynamics. They represent, in fact, the *ultimate* level of generalization, one that has hitherto remained hidden to physics and to empirical philosophy in general. It is hopefully one of the achievements of this work to bring this most fundamental level of all (the level of logical necessity) to light.

This brings me to my next argument concerning the deep identity between information theory and thermodynamics. This deep identity, as has been observed, is *itself* built upon the precise identity between thermodynamics and geometry which was first identified by Ludwig Boltzmann. By generalizing from his initial insight it is in fact possible to recast the three laws of thermodynamics in terms of symmetry conservation and symmetry breaking. Thus the first law implies that symmetry cannot be created or destroyed. The second law implies that a progressive deterioration in symmetry over the life of the universe (i.e. entropy) should be balanced (in a closed universe) by the progressive recovery of that lost symmetry. The third law implies the unattainable nature of perfect symmetry. These (which are in reality simply a particular application of Noether's theorem) we might call the three laws of symmetry.

Since we have established that the first two laws of thermodynamics are directly translatable into the language of information theory, what of the third law of thermodynamics? Well, this question has already been answered. By demonstrating that the first engineering function of the universe is, in reality, a generalized form of the third law of thermodynamics, and by demonstrating that the first engineering function can be restated in terms of information I have, in effect, already iterated the third law of information theory, which is that; *a system can never encode zero or infinite information.*

30. Indeterministic Order.

The interpretation of quantum mechanics in isolation is a simple and unequivocal one; indeterminacy. But an interpretation of physics *as a whole* incorporating the implications of classical physics as well (notably thermodynamics) is equally unequivocal; ***Indeterministic order***. Which is to say; that order emerges on the classical scale out of indeterminate processes on the quantum scale.

But this interpretation is based on two assumptions; a closed universe and the conservation of entropy – as discussed earlier. These granted the interpretation of physics as a whole (as distinct from quantum mechanics in splendid isolation) as pointing to a universe of indeterminate order is both complete and indisputable.[39]

But if entropy is *not* conserved the universe *will* become increasingly disordered and chaotic until these properties reach a maximum at thermal equilibrium (i.e. the heat death of the universe). Under this model the correct interpretation of quantum mechanics in the context of classical physics is equally explicit; ***Indeterministic chaos***. These are the *only* two feasible interpretations of physics as a whole. At present it is (de-facto) the latter interpretation which holds sway and is under challenge in this work.

It should be clear in any case that given the relevance of issues such as the fate of the universe and the conservation or otherwise of energy and entropy, that equal significance must be given to classical physics in any comprehensive interpretation of physics as a whole. The almost complete neglect of this approach (which alone yields a true interpretation of physics) can perhaps be attributed to the fact that, in recent years, we have allowed ourselves to be seduced by the tantalizing idea of a so called "final theory". Such a theory may well already exist in the proto-typical form of so called "M-theory", but it is vital to recognize that any such theory will be a final theory of *quantum mechanics* and not of physics as a whole. Anyone who thinks that the discovery of such a theory (which will in effect give us a comprehensive understanding of all possible quantum-scale interactions) will render classical physics redundant (as theory let alone as practice) is simply being naive. The truth is that such a theory, which will serve to clear up problems surrounding the gravitational interaction, will most likely be anti-climactic in its effect and have very few practical consequences.

It is even possible to argue that, given the validity of Gödel's incompleteness theorem, any final physical theory of the level of generality aimed at by M-theory is inherently undiscoverable. But, since "incompleteness" will be built into the foundations of any such theory (in the form of the uncertainty principle) this argument for what amounts to so called "cognitive closure" is not valid. This is simply because the *epistemological* significance of indeterminacy and incompleteness is identical. Insufficiency of data from cosmology and from high energy physics is therefore a more likely stumbling block to the final formulation of such a theory in the future.

At any rate, Indeterministic order, suggesting as it does that God plays with loaded dice, ironically implies that both Niels Bohr and Albert Einstein were ultimately correct in their arguments concerning

---

[39] Elimination of the need for determinism as a first order phenomenon is simultaneously an elimination of the need for a sizable portion of Kant's system which sought to account for the phenomenon of objectivity by supplying arguments for the apriori status of causality. Indeed it seems to me that the apriori arguments for symmetry and symmetry breaking (which together account for the phenomenon of intersubjectivity and orderliness, including causality as a limit case) are a good deal clearer. The arguments for perfect symmetry (in essence for absolute nothingness) scarcely need making (since the puzzle is not why *nothing* exists but rather why *anything* exists), whilst the arguments for symmetry *breaking* (i.e. for the generation of phenomena) are encapsulated in the expression of the *first engineering function* and are quite clearly apriori in nature since, although derivable from thermodynamics, they are ultimately *logical* rather than physical in nature and hence ipso-facto apriori.

As such we say that the arguments deployed in part one of this work represent, amongst other things, a comprehensive renovation of Kantian apriorism, with recourse *not* to the categories and the forms of intuition (which are not so much wrong in their drift as they are irrelevant) but rather, at a more fundamental, clear and objective level, to the two engineering functions (in effect, the principle of symmetry and the principle of symmetry breaking) that are deducible from logic and from modern physics alike.

the interpretation of science. Their argument seems to have been at cross purposes since Bohr was correctly identifying the epistemological significance of quantum mechanics in isolation (i.e. indeterminacy) whereas Einstein was (I think correctly) discussing the likely epistemological significance of science *as a whole* (i.e. order). The correct fusion of their contrary perspectives therefore, which is only apparent to us now, must be (and indeed is) *Indeterministic order.*

31. The Correspondence Principle.

What is perhaps most important about the interpretation of physics as a whole is that it implies (as does our analysis of the limits to a "theory of everything") the existence of a fundamental *complementarity* between classical and quantum physics. This is ultimately because quantum physics, although it is the more general or universal of the two systems (with classical physics constituting a special or limit case with regards to it), cannot practically be applied at the classical scale. Such an application would in any case be superfluous since the results it gave would "average out" (i.e. "smooth out") the net quantum randomness to yield virtually the *same* answer less laboriously arrived at via the classical (and in effect heuristic) method. Consequently a true "final theory of everything" must comprise a quantum theory coupled with classical physics as a useful approximation at the classical scale. The only qualifier to this picture is that, in the event of any conflict of interpretation, quantum theory, being the more general form of the final theory, would have epistemological priority over classical physics in every case. But this is more a technical issue for philosophers than for working scientists. Certainly, linguistic incommensurability between the classical and the quantum paradigms ceases to be a significant problem on such a reading.

The basics of this interpretation indeed already exist in the form of Niels Bohr's well known "correspondence principle", the so called "classical" or "correspondence" limit being reached where quantum numbers are particularly large, i.e. excited far above their ground states. A complementary definition is that the classical limit is reached where quanta are large in number. Though sometimes criticized as heuristic there is in reality no reason to doubt its enduring validity given the above caveats. Indeed, the theory is probably Bohr's only contribution to the *interpretation* of quantum mechanics which is of lasting value. As such it is undoubtedly one of the central principles of empirical philosophy in its final form and a much under-appreciated one at that. Without it incommensurability would be a serious problem.

The crucial link between the two paradigms is further demonstrated by Boltzmann's statistical reinterpretation of the second law of thermodynamics, a reinterpretation at once confirmed and explained by the subsequent discovery of quantum mechanics and the uncertainty principle. Indeed, the reinterpretation of the second law of thermodynamics coupled with the later discovery of quantum mechanics necessitates a reinterpretation of *all three* laws of thermodynamics along similar statistical lines. Nevertheless, despite being probabilistic and with an underlying quantum mechanical *modus operandi* the laws of thermodynamics are still unequivocally laws of the classical scale that illuminate scientifically and epistemologically important facts that we would not otherwise be apprised of were we to rely solely on a quantum theory. It is therefore this evident link between the two scales (and indeed paradigms) of science which contradicts those who deny the existence of any basis at all for the unity of quantum and classical physics. It also confirms our interpretation of the correspondence principle; that it is the valid model for the *complementarity* of the two paradigms or scales of physics. In any case, without the correspondence principle or something similar the interpretation of physics *as a whole* (i.e. as demonstrating the existence of indeterministic order/chaos) becomes impossible.

32. Concerning Incommensurability.

Our analysis of Boltzmann's reinterpretation of the second law of thermodynamics also enables us to deduce that classical physics itself is not truly deterministic at a fundamental level. This is an important point to make if only as a counter to certain aspects of the so called "incommensurability argument" of Thomas Kuhn, an argument which, as we have just seen, is occasionally adduced (mistakenly) against the fundamental validity of the correspondence principle. But if this principle were not valid scientists would be reduced to making classical scale predictions out of vast aggregates of quantum scale data, which is obviously absurd. Indeed, it is precisely *because* of incommensurability between quantum and classical physics that the correspondence principle has been postulated.

Fortunately, Boltzmann's statistical reinterpretation of thermodynamics, coupled with the subsequently uncovered link to quantum mechanics enables us to assert the validity of the correspondence principle with some confidence, not least because it implies that thermodynamics and with it the rest of classical physics cannot legitimately be given a simple deterministic interpretation after all.

Furthermore, the technical evidence for classical indeterminacy as supplied by Maxwell and Boltzmann's discovery of statistical mechanics (and also by Poincare's analysis of the so called three bodies problem) corroborates David Hume's earlier and more general discovery of classical indeterminacy as eloquently described in the groundbreaking "Enquiry Concerning Human Understanding".

Hume's reading of the problem of induction coupled with the subsequent discovery of the statistical mechanics refutes the alleged dichotomy between classical and quantum physics concerning the issue of causality. It consequently suggests that the correspondence principle is indeed a valid model for the *epistemological* commensurability of the two paradigms of science. As such, the perceived incommensurability of *language* between classical and quantum physics is not as significant as we might suppose since the nature of the phenomena under discussion in each case is also different.

A different issue concerns incommensurability between theories *within* classical physics such as that which began to develop between electrodynamics and Newtonian mechanics towards the end of the nineteenth century. But the development of crises within physics, of which this is a classic example, often point to resolution at a deeper, more comprehensive level of theory. In this case the resolution occurred through the development of quantum and relativity theory. Interestingly, the overturning of Newtonian mechanics (whose incommensurability with electrodynamics was not at first apparent) was merely incidental to the solving of a problem in electrodynamics (concerning ether drag). It was not due to anomalies in Newton's theory (although these were recognized) and hence not directly due to Popperian falsification. Einstein's extension of the theory of relativity to the problems of gravity also enabled science to solve certain problems concerning the nature of time and space that had previously been considered problems of metaphysics and had been unsatisfactorily handled by Newton.

Thus, whilst the shift from Aristotelian to Classical physics indicates a paradigm *shift* as from a less to a more *rigorous* concept of scientific methodology, that between Classical and Quantum physics (which share the same dedication to experimental rigour) is more accurately characterized as a paradigm *expansion*. Thomas Kuhn's attempt to argue that the two most profound shifts in the history of science are somehow epistemologically the same is therefore fundamentally misguided, as is his attempt to suggest that a profound and unbridgeable gulf separates classical and quantum physics whereas it is infact more accurate to say that quantum physics retains classical physics as a *special case.*

Most of the lesser "shifts" discussed by Kuhn are again of a different kind, indicating Kuhn's well known failure to make adequate distinctions. They usually represent comparatively trivial instances of incorrect or incomplete empirical hypotheses being displaced by more accurate and complete ones. This is obviously the normal way of science except that Kuhn chooses to characterize the displacement of hypothesis A by hypothesis B as a *conceptual* shift, which, I suppose, technically it is.

Furthermore, anomalies which lead to minor corrections and those which lead to major corrections are discovered and treated *identically* from a methodological point of view, thus making Kuhn's distinction between "normal" and "revolutionary" science dubious, at least at the most fundamental level. The full significance of anomalies, after all, usually only becomes apparent with hindsight, which is why scientists tend to be conservative in the ways Kuhn indicates, retaining theories, even in the face of anomalies, until such time as better, more general and more precise hypotheses arrive to replace them.

33. The Problem of Causation.

The question that arises at this point concerns the status of causality, which is undermined by quantum theory at an empirical level and by Hume's identification of the problem of induction at a more fundamental epistemological level.

Although Kant is doubtless correct in identifying causality and the other categories as necessary preconditions for the very existence of consciousness, it does not therefore follow that causality *itself* is fundamental. This is because there are no reasons for believing that consciousness and selfhood (both of which are transitory, composite phenomena) are fundamental either, notwithstanding the epistemology of Descartes or of Kant.

What then accounts for the appearance of causality at the classical limit? Our analysis hitherto suggests that causality, like time and space, is only a second order phenomenon generated by the spontaneous breaking of symmetry at the radiation origin. In this case invariance (the second engineering function in other words) is the more fundamental or general phenomenon, which space, time and hence causality itself are (perhaps inevitable, or "apriori") by-products of.

That the Kantian categories as well as time and space are consequential to the two engineering functions would seem to lend the engineering functions themselves an *apriori* character, implying the superfluity of Kant's system, except, perhaps, as a special case of the neo-rationalist model described in this work. But complementing this observation we must also note the fact that both functions are also derived aposteriori as a result of the empirical method. They are *not* quasi-metaphysical abstractions from experience in the way that the categories are. The fact that the engineering functions also make intuitive sense does not make them any less empirical in nature. Similarly, the recognition that causality *itself* is a limit phenomenon is inferred from the consideration of empirical data.

Nevertheless, it is precisely *because* symmetry is such a fundamental principle applying, unlike causality, even at the quantum scale, that classical physics is able to "work". Our fear concerning the loss of causality (rightly felt by Kant) is thus abated when we realize (as Kant was in no position to do) that causality is but a special case of this far more general principle. And so the "unreasonable" effectiveness of classical physics is preserved *without* the need to hypothesize (as Kuhn assumed) a consequent incommensurability with quantum theory. Commensurability, coupled with a convincing account of the effectiveness of classical physics is only possible if we are prepared (as the data suggests we must) to reinterpret causality as the limit phenomenon of a still more general principle.

For even quanta, which do not obey the laws of cause and effect (since these are "laws" of the limit) *are* obliged to obey the principle of invariance as expressed in the form of the various conservation laws of classical physics. As a consequence of this the appearance of causality and hence the effectiveness of the "laws" of physics is maintained at the classical scale. Thus, just as classical physics is displaced by quantum physics as the more fundamental paradigm of science (becoming in effect a limit phenomenon of the latter) so causality is displaced by invariance as the more fundamental principle associated with this more inclusive paradigm. Kuhn's pessimistic model of warring paradigms is in effect displaced by Einstein's less famous model of expanding and inclusive ones.

It is of course reasonable to ask the question "Why symmetry?" But this in turn reduces to the question "Why perfect order?" which was answered at the end of section fourteen. For in the end symmetry requires *no* explanation, it is the *deviation* from perfect symmetry which is the true mystery. This indeed is the principle reason for considering the second engineering function – the principle of net invariance – to be apriori, like the first engineering function. If there were *not* net perfect order we would find the source of entropy inexplicable. Therefore Curie's principle of symmetry, amongst other things, depends on the validity of the principle of net invariance. Thus the mere existence of entropy proves that the second engineering function is a necessary assumption.

And this is perhaps the profoundest argument of all in favour of the radical "conservation of entropy" hypothesis presented earlier in this work. Invariance is, after all, nothing other than the principle of parsimony by another name. Therefore if entropy is *not* conserved (as is the current orthodoxy) then this amounts to its being given away for nothing, which is poor accounting on the part of the universe and not very probable in my opinion. More damningly, un-conserved entropy implies non-conservation of energy as well, since the two concepts are closely allied. After all, if the universe goes on expanding forever, or else contracts into a chaotic blob we also have a lot of left over and rather useless energy whose existence no-one can account for and all because we were not willing to countenance the possibility that entropy might be conserved (like everything else is believed to be) as well.

34. The Physical Significance of Indeterminate Order.

If the net entropy of the universe equals zero, as I propose it must, then "chaos" or disorder are revealed to be anthropocentric concepts merely relating to the inherent unpredictability (as distinct from disorderliness) of complex phenomena.

An example of this phenomenon (of what I call "indeterminate order") is the case of atomic decay. Whilst individual instances of atomic decay *are* inherently unpredictable (ultimately because of spontaneous symmetry breaking and the uncertainty principle) they nevertheless display a collective orderliness as exampled by the fact that the half life of any element is readily determinable. This orderliness which emerges at the classical limit out of quantum indeterminacy is due (according to the theory of indeterminate order) to the fact that quantum phenomena are nevertheless obliged to obey classical conservation laws, thereby conforming to the macro-principle of symmetry.

As a result of this the progression of disorder in a system is paradoxically predictable (by means of Boltzmann's equation and the statistical mechanics) – something mimicked by the graphics of so called "chaos theory", the famous sets and attractors.

Surprisingly even mathematics furnishes a fascinating example of the phenomenon of indeterminate order in the case of the distribution of primes, a problem described by David Hilbert as the most important problem "not only in mathematics, but absolutely the most important."

Although the distribution of individual prime numbers appears to be inherently unpredictable nevertheless the prime number theorem together with Riemann's treatment of the zeta-function allows us to determine precisely *how many* primes appear in any given number (for instance, the number 10 contains four primes; 2, 3, 5 and 7). This discovery indicates that though the distribution of primes is unpredictable it is not therefore chaotic since the graph of the distribution of primes rises smoothly even in spite of an unpredictable distribution. The parallel with the approximate predictability of the half life of atomic elements, notwithstanding the inherent unpredictability of atomic decay, is certainly quite striking, and, as we shall see, points to a common underlying link with the progression of entropy, itself a paradoxically orderly affair. Indeed the whole issue of entropy explodes the simplistic distinction between chaos and disorder.

Nevertheless, the simplest and most illuminating analogy is with the tossing of a coin. Although the result of each individual coin toss is inherently unpredictable, nevertheless the *aggregate* of all coin tosses is remarkably predictable within a given margin of error defined by standard deviation. And this margin of error (as a proportion of the number of coin tosses) shrinks as the number of coin tosses increases, making the distribution increasingly even. In essence the prime number theorem and its associated graph (the graph of the function Pi (n)) tell us *exactly* the same thing about primes. Their individual distribution is inherently unpredictable (as with individual tosses of the coin) but the total number of primes less than a given number (n) is predictable within a margin of error that diminishes (as a proportion of (n)) as (n) grows larger, leading to the increasing smoothness of the graph of Pi(n).

In fact, the number of primes in existence appears to be the absolute minimum that is required to fulfill the condition that all other natural numbers be the product of primes. This is why primes are sometimes referred to as "atomic numbers" ("the jewels in the crown of number theory" as Gauss poetically described them) whilst the remaining natural numbers are called "composite numbers". One could equally well call primes "analytical numbers" and composites "synthetical numbers".[40]

In the mystery of the distribution of primes we therefore detect not only the presence of indeterminate order but also the operation of the principle of parsimony. What might be called the structure of number theory is indeed nothing more than a highly rarefied form of geometry, the geometry of numbers so to speak. Number theory is therefore simply the analysis of aspects of this structure – what might be called

---

[40] Indeed, another interesting analogy is with primary and secondary colours in colour theory. See the appendix to this work for a fuller treatment of the issue of prime numbers and entropy.

the pure internal geometry of numbers themselves. And this is why indeterminate order (the unpredictable distribution of the primes) and the principle of parsimony show their ghostly hands in the most important problems of pure mathematics as well.

Another example of indeterminate order and parsimony concerns the previously discussed issue of the physical constants. The apparently random nature of these constants in many ways represents as big a mystery for the physics community as does the issue of primes for mathematicians. The fact that these constants cannot generally be derived from theory appears to be an integral feature of nature and is certainly not the result of the absence of some "theory of everything". Nevertheless, despite this randomness the physical constants obviously contribute in a fundamental way to the overall order of the universe as a whole, just as primes contribute to the construction of the mathematical continuum.

These various instances of indeterminate order (and there are countless more of them) are certainly interconnected. And a common denominator for all of them is that they prove susceptible to a description in terms of "complex numbers". And complex numbers are of course composed out of so called "imaginary numbers", numbers that are (metaphorically speaking) beyond the "visible spectrum" of the more familiar "real" numbers but are no less (or more) real for all that. But, being irrational, they seem to inevitably generate indeterminacy and unpredictability.

Imaginary numbers are irrational in that their expansion in any particular base never ends and never enters a periodic pattern. Therefore, the existence of irrational numbers (which include transcendental numbers as a special case) foreshadows Gödel's époque making proof of incompleteness in a non trivial way, explaining its inevitability to some extent. Transcendental numbers were after all the great scandal of ancient Greek mathematics (although they did not cause the same level of anxiety to ancient Indian mathematicians such as Bramagupta)[41]. As such, like Gödel's powerful proof, irrational numbers perhaps point to the inductive foundations of mathematical reasoning.

And this is of particular significance to physics given that quantum mechanics, as a comprehensive description of nature, has been found to be embedded in a Hilbert space whose very coordinates are defined by complex numbers. This non-local or "phase" space thus constitutes the geometrical foundation for quantum mechanics, just as Riemannian space does for Einsteinian mechanics and Euclidean space does for Newtonian mechanics. And just as quantum mechanics retains Einsteinian (and hence Newtonian) mechanics as a special case, so Hilbert space retains Riemannian (and hence Euclidean) space as a special case as well. Consequently, the geometry of Hilbert space is the most general geometry of all.

And this incidentally explains why so called "loop quantum gravity" theory (the attempt to make quantum gravity conform to the geometrical schema defined by Einstein, thereby making it "background independent") is conceptually flawed. The fact is that quantum phenomena attain so called "background independence" simply by virtue of being described in terms of their own internal phase space – which is the geometry of Hilbert and Cantor. Thus the attempt to describe quantum phenomena in terms of classical or Riemannian space represents a step backwards and a failure to understand the nature and indeed the necessity of the new trans-finite geometry.

More familiar examples of indeterminate order are also detectable in the phenomenon of so called "self-organization". Indeed, self-organization is properly interpretable *as* a phenomenon of indeterminate order, indeterminate order in action so to speak. Thus the evolution of cities, nations, economies, cultures

---

[41] The classical problem of squaring the circle is perhaps the earliest allusion to the problem posed by transcendental numbers to the rational integrity of mathematics. This is because this problem is identical to the more serious one of finding the *area* of the circle (i.e. the question of the final value of Pi). It is because the latter is not possible that the former is not either. If one says "oh, but the area of a circle is πr^2" then one can as easily square the circle by finding $\sqrt{\pi} r^2$. But this is of course impossible because pi is a transcendental number, a fact finally proven in 1882 by Von Lindemann. This problem is consequently an early and illuminating example of mathematical uncertainty.

Indeed, we can even conclude from this paradox that there is no such thing as an ideal mathematical circle in the Platonic sense either. The proof of the transcendental nature of Pi is therefore incidentally also a disproof of Plato's theory of forms. Not only are all *actual* circles merely approximations, but the mere *idea* of a perfect mathematical circle is absurd as well. There are no real or ideal circles and hence nothing to square.

and even life itself (in which a form of overall order emerges out of the sum of unpredictable, though often intentional, acts and events), all represent prime examples of self-organization, i.e. indeterminate order in action. This is because these acts and events generate continuous positive feed-back loops and hence unpredictability (indeterminacy) combined with emergent order. All of which is what should be expected in a self-organizing universe devoid of design, but driven into existence by the various tensions existing between underlying logical and thermodynamical necessities.

35. Deterministic Chaos.

If entropy is not conserved, however, the principle of net invariance (the second engineering function) is invalidated and the correct interpretation of physics as a whole defaults to the more familiar one of *indeterministic disorder* or chaos. But if this is the case then, as indicated earlier, the question of the origin of entropy would represent itself and Curie's principle of symmetry would be invalidated.

Despite the inevitability of this interpretation it is a remarkable fact that a rival interpretation of science as a whole also exists; that of "deterministic chaos" as supplied by so called "chaos theory".

This alternative interpretation stems from treating chaos theory, which is primarily a contribution to applied mathematics (with some marginal implications for the sciences) as a *new* science. It is the result of a misguided self-aggrandizement in other words. Nevertheless, the attempt to apply an interpretation from a highly synthetic "field" of mathematics to science as a whole is inherently wrong and leads to conclusions which are embarrassingly simplistic.

As the graphical application of non-linear mathematics and complex numbers chaos theory is usually interpreted as exhibiting deterministic chaos, which, given our earlier analysis, cannot be the correct interpretation of physics and hence of science as a whole. This is because one cannot artificially separate off the classical from the quantum scale and proceed to interpret science as if the latter did not exist (chaos theorists, it seems, will always get round to dealing with quantum mechanics at a later date).

Another major reason for the comparative conservativism of chaos theory and also for its inherently limited practicality (notwithstanding the numerous claims to the contrary on its behalf) is the absence of any truly new mathematics at the heart of the theory. It is no accident, incidentally, that the rise of chaos theory coincides with that of the modern computer. This coincidence points to the primarily applied (as opposed to theoretical) significance of chaos theory, a theory fundamentally misunderstood by its foremost proponents.

Such mathematical tools as chaos theory *does* make limited use of are already to be found in play, in a far more sophisticated form, especially in quantum physics. From this point of view the uber equation of chaos theory should really be the Schrödinger wave-function, except that this equation leads to a different interpretation of science than the one they hold so dear. By contrast the adapted Napier-Stokes equations of a strange attractor or the simple iterations in the complex plane of a Mandelbrot or a Julia set are comparative childs-play.

The primary value of the sets and attractors of chaos theory, as classical objects, is that they allow us to see something of what these tools "look like" when applied at the classical scale – and what they resemble is a combination of turbulence and fractal self-similarity. The Mandelbrot set in particular demonstrates that an unpredictable system – such as the universe itself is – can nevertheless display perfect order combined with a literally unlimited richness of structure – a fact which stems from the elegant use of phase space in constructing the set. Although the Mandelbrot set has no physical significance whatsoever it still probably ranks as the single most outstanding discovery of chaos theory so far and one of the greatest discoveries of twentieth century mathematics.

Nevertheless, despite repeated insinuations to the contrary, chaos theory cannot significantly help us to predict turbulence since turbulence is the evolving product of systemic entropy and so to be able to predict it would amount to contravening the uncertainty principle and hence also the second law of thermodynamics.

Unlike the equations of chaos theory however, the wave-function is *not* deterministic since it only predicts probabilities. And this is the key difference separating the deterministic mathematics of chaos theory from the more elaborate probabilistic mathematics of quantum mechanics. It points to the fundamental weakness of chaos theory as an interpretation of science; to wit the assumption that initial conditions are, even in principle, determinable. This is an assumption disposed of nearly a century ago by quantum mechanics and, in particular, by the existence of the uncertainty principle.

As such the true interpretation of physics can *only* be an indeterministic one, thereby contradicting the underlying assumption of chaos theory in a definitive way. The famous "butterfly effect" is therefore a quantum rather than a deterministic phenomenon in its origins. Sensitive dependence remains, but initial conditions are shown to be a classical idealization of the theory.

Indeed this error is not even new in the history of physics and may date back as far as the time of Pierre de Laplace in the eighteenth century. Certainly, by the time of Poincare and the statistical mechanics a century later the interpretation of physics was indisputably one of deterministic chaos, informed as it was by the determinism of Newtonian mechanics on the one hand and by the discovery of the second law of thermodynamics on the other. Modern chaos theorists have unwittingly reverted back to this purely classical paradigm, which was abolished, in science at least, by the discovery of quantum mechanics. Consequently we may say, in Wolfgang Pauli's immortal phrase, that it is "an insufficiently crazy theory".

This is perhaps why chaos theorists seem incapable of accepting that an acausal sub-strata (described by quantum mechanics) underpins their deterministic and classical models, or else they dismiss any reference to this inconvenient fact as "trivial" and so not worthy of the kind of consideration that I have given it here. For recognition of this basic fact renders their fundamental and much treasured "discovery" of deterministic chaos (which was never even remotely original) irrelevant.

36. Why the Anthropic Principle is not even Wrong.

The Anthropic principle appears to be metaphysical since there is no conceivable means of testing its various assertions, it is therefore "not even wrong" in Wolfgang Pauli's other immortal phrase. And since the universe is the product of the decay of a perfect or near perfect state of order (even inflationists agree here) it should be of less surprise that it happens to exhibit enormous richness of structure. Furthermore no one is truly in a position to say whether alternative universes may or may not be capable of generating some form of organic life, no matter what statistics they may adduce in their support.

What one *can* say is that in any universe where organic life *does* emerge the evolution of consciousness – a useful tool for the survival of life – is likely to be an eventual concomitant. On earth alone brain evolution has taken place independently in a number of completely different phyla ranging from arthropods to chordates.

On this simpler and more rational reading *consciousness* (like life itself) is simply an *emergent consequence of complexity* and, as we have seen, complexity is a product of the two apriori (and empirical) engineering functions of the universe.

The principle of evolution, as we will see in part two is of much wider application than is commonly assumed. Even the parameters of our universe, as determined by the physical constants seem to be the by-product of natural selection, which is why the physical constants can only be known from observation and not derived from any theoretical structure. What is driving the selection of the physical constants however is the second engineering function, the requirement that all symmetries be conserved. We may therefore assume that the physical constants are self organizing with a view to this one global end. And any universe displaying this sort of net global order could conceivably support the evolution of life as well, at least in some form. The point is we cannot certainly know one way or the other, hence the designation of the anthropic principle as unempirical, since its validity does not follow from any of our current empirical formalisms and seems to point instead to a lack of knowledge of underlying empirical principles, notably of what I term the two engineering functions of the universe.

What is important from the point of view of the universe (so to speak) is not that life or consciousness should emerge, but rather that, notwithstanding the dictates of the first engineering function, symmetries should nevertheless be conserved – i.e. no wastage should occur. And if tremendous richness of structure, including life itself, should be the by-product of this interaction of the two engineering functions (as seems to be the case) it really should be no cause for surprise, given what we already know concerning their implications and their necessity.

I am not sure that the principle of net invariance *can* be used as an *empirical* alternative to the conveniently amorphous Anthropic principle however, since, though it *does* explain the phenomenon of emergent order (and hence by implication the phenomena of life and consciousness as well) it does *not* explain it in a way that satisfies the quasi-religious (or indeed economic) yearnings of those who proselytize on behalf of the Anthropic principle (including most authors of popular accounts of modern physics) and who therefore purport to detect mysteries and improbabilities (notably the so called "tuning problem") that are quite likely chimerical in nature and hence inherently inexplicable.

37. The Measurement Problem.

The epistemological and ontological significance of quantum mechanics (which is the most comprehensive physical theory) is fully expressible in three concepts; indeterminacy, acausality and non-locality. Bearing this in mind the famous "measurement problem" concerning what causes the so called "collapse" of the Schrödinger wave-function is akin to asking; "what collapses the racing odds?" It is a foolish question and the measurement problem is a non-problem. Quantum processes are indeterminate therefore there *can* be nothing to cause the "collapse". Searching for solutions to this and other aspects of the measurement problem therefore represents nothing more than a hankering for old style causal thinking, thinking which is no longer sustainable in view of the uncertainty principle and other aspects of quantum mechanics. [42]

Though quantum states are undefined prior to measurement they nevertheless obey thermodynamical principles of invariance, principles which simply express themselves through and indeed *as* the disposition of all quantum phenomena. Measurements are therefore approximate descriptions of certain aspects of the overall invariant entity that is the hyper-atom. They are at once *explorations of* and *contributions to* this invariance. This is the paradox pointed to by the existence of the uncertainty principle.

Symmetry is therefore the physical principle which allows us to reject causality and metaphysics (in line with Hume) and yet *at the same time* to retain rationality and order and hence the validity of objective judgment (in line with Kant). It therefore represents the coming of age of physics and hence of logical empiricism as a comprehensive philosophy no longer in need of any external metaphysical support. This is because the principle of symmetry effectively displaces (or rather *marginalizes*) the causal principle (which Kant's system was created in order to defend) and yet counteracts the agnostic implications of quantum indeterminacy as well. It is therefore of vital epistemological significance to any comprehensive empirical philosophical system.

As we have seen even classical physics relies on the principle of symmetry at least as much as it does on the principle of causality since it consists primarily of assertions that certain quantities must always be conserved or else that given a certain set of circumstances certain epiphenomena will always be observed. Indeed, this latter assertion (which is the principle of causality) is merely a restatement in other terms of the more fundamental principle of symmetry.

Quantum mechanics therefore allows us to resolve fundamental problems of ontology and epistemology in a decisive and permanent way. This is its true significance for empirical philosophy.

---

[42] Recent discussion surrounding so called "quantum determinism", (representing an implied *alternative* interpretation of quantum mechanics), indicates a profound misunderstanding of the meaning of the square of the wave-function. Since the wave-function assigns a given *probability* to every conceivable event one can therefore say that it predicts everything that ever happens. However, to interpret this as a species of "determinism" is absurd and only succeeds in vitiating the word of all sensible meaning.

38. The Nihilistic Interpretation of Quantum Mechanics.

But how are we to arrive at an "invariant" interpretation of quantum mechanics? Such an achievement would be no small thing and would resolve on a permanent basis many problems of ontology and epistemology. Its significance would therefore extend far beyond that of physics itself or even the philosophy of science. This is because, as mentioned repeatedly, quantum mechanics is the most general physical theory of all. Therefore if we are to base ontology and epistemology on empirical foundations then the interpretation of quantum mechanics will be of decisive significance.

The key to a reliable interpretation lies in establishing its formal foundations. These foundations must be unimpeachable and hence universally accepted amongst authorities if we are to build a successful analysis based upon them and if we are to eliminate erroneous interpretations with reference to them. They shall form, as it were, the *premises* of all our arguments on this topic.

Foundations meeting these strict criteria are, I believe, only three in number, to wit; Max Born's statistical interpretation of the Schrödinger wave-function , Werner Heisenberg's uncertainty principle and John Bell's inequality theorem. Together, I believe, these mathematical formalisms virtually *enforce* the correct interpretation of quantum mechanics. Furthermore, any interpretation which strays from their implications or adduces arguments based on alternative foundations, is ipso-facto wrong. As such these three formalisms perform a crucial *eliminative* function as well.

Born's contribution lies in demonstrating that the *square* of the Schrödinger wave-function (the central formalism of quantum mechanics) is a probability amplitude and *not* a literal description of quantum phenomena as first assumed. It is from this discovery as well as from that of the Heisenberg uncertainty relations that the basic solidity of the orthodox Copenhagen interpretation of quantum mechanics (C.H.I.) entirely stems.

Incidentally, the main contemporary rival to the C.H.I. (Richard Feynman's "path integral" interpretation) can be seen as representing a partial reversion to the literalistic interpretation that was discredited by Born's discovery. It is an upgraded version of the literal interpretation, designed to take the statistical interpretation into account as well. As a result of the attempt to derive a visual picture (as distinct from a purely mathematical one) the sum over histories interpretation has a distinctly metaphysical feel to it. The truth is that the path integral approach is simply *another* mathematical model of quantum mechanics and so we are no more justified in building a physical picture from it than we are from any of the many other alternative expressions of quantum mechanics. Such pictures may be consoling, but they are never justified.

Ironically Feynman himself seems to have understood something of this since in stray comments he increasingly came to espouse an *instrumentalist* interpretation of quantum mechanics which he dubbed the "null interpretation". Unfortunately this refinement (in effect a reversion to Born's orthodox position) is not what he is remembered for today.

In echoing C.H.I. the "null interpretation" also implies the validity of non-locality and so, like C.H.I. points squarely to nihilism as the underlying ontological significance of quantum mechanics. Indeed all true interpretations of quantum mechanics, since they must be in line with the three aforementioned formalisms, point inescapably in this direction.

In practice there *are* only three possible generic interpretations of quantum mechanics; the nihilistic, the literalistic (and its refinements) and the hidden variablistic (in effect holistic). Born's instrumental interpretation of the wave-function rules out literalistic or pictorial interpretations, whereas experimental violations of Bell's inequality rule out hidden variable interpretations of the sort favoured by Bell himself. This leaves only nihilistic and instrumentalist interpretations of the sort implied by C.H.I. and the null interpretation.

Although non-locality has been established with reference to Bell's theorem (and it is also strongly implied by our interpretation of the uncertainty principle supplied in section six) the counter intuitive nature of ontological nihilism is so great that the proper ontological significance of quantum mechanics has always been missed (and is still missed) by physicists and philosophers alike down to the present day. Nevertheless, non-locality is what we would expect to find if onto-nihilism is infact true. It is, indeed, a less formal way of expressing this most fundamental of all ontological stances.

39. Wonderful, Wonderful Copenhagen.

Although quantum mechanics offers no picture of underlying phenomena quantum *theory*, rather surprisingly, is a different matter. According to quantum theory particles are simply *"discrete waves"*, but with highly compressed wavelengths. This account explains wave-like behaviour, but it also explains the particle-like behaviour of quanta as well, as in well known phenomena such as the photo-electric effect. Since all particles are defined in wave-like terms of frequency it follows that they must have a wave-length and hence dimensionality as well. As such they are in reality discrete *waves* albeit at the high frequency end of the spectrum. The true interpretation of phenomena such as the photo-electric effect is therefore that electro-magnetic waves at very high frequencies behave *like* particles but are nevertheless still waves. This leads us to the general insight which is that what we call particles are in reality special or limiting cases of discrete waves – a crucial interpretation which obviously eliminates the orthodox concept of the wave-particle duality, thereby necessitating an overhaul of the standard Copenhagen interpretation along the more simplified and economical lines described in the preceding section.

Indeed, this revision finds powerful support from recent developments in string theory based as it is on the fact that quantum theory sets a natural minimal limit, defined by the Planck length. This in effect rules out particles as physically meaningful concepts entailing thereby an abandonment of Bohr's concept of "complementarity" as well. This concept, along with the whole doctrine of wave-particle duality, should be replaced by the analysis of the universality of discrete waves (which retains the relevant elements of *both* models) given above.

One might think that this "discrete wave" interpretation of quantum theory would be at odds with string theory as well, but this is only superficially the case. This is because string theory reinterprets the particle concept to the effect that *all* "particles" are treated as vibrating "open" strings. And what is a "vibrating open string" if not a discrete wave? The *one* exception to this valuable overhaul of the particle concept offered by string theory is the mysterious graviton. Since the graviton exhibits the unique characteristic of possessing *two* units of spin it is interpreted by string theory as being a *closed* string.

But, as far as the current standard model of all quantum interactions (excluding gravity) is concerned it is indeed the case that, as Erwin Schrödinger once remarked, "All is waves". And any account of gravity in terms of quantum mechanics should yield a wave equation of a form similar to the Schrödinger equation, in effect describing nothing physical, but a probability amplitude instead.

In addition to the overhaul of the Copenhagen interpretation implied by quantum theory a nihilistic interpretation of quantum mechanics also occasions an observation concerning the realist/instrumentalist dispute which is a mainstay of the philosophy of science. Although quantum mechanics is purely instrumental it is nevertheless the case that quantum theory, as we have just mentioned, gives a discrete wave or "string" picture of quantum phenomena. Nevertheless, since quantum mechanics is ultimately indicative of non-locality it can reasonably be argued that even quantum mechanics, the acme of instrumentalist interpretations for the last eighty years, is a "realist" theory since, by giving us *no picture at all* of the "underlying" reality or phenomenon it is in practice giving us the *true* (i.e. non-local) picture.

Thus the distinction between quantum mechanics as an *instrument* and quantum mechanics as *ontology*, though a valid one, is not mutually exclusive as hitherto thought. By pointing to transience and universal relativity, quantum mechanics is indeed the mechanics of inessence. As such no picture of phenomena is to be expected, and, in line with the null or *instrumentalist* interpretation first offered by Max Born, none is given. But, again, only an onto-nihilistic interpretation of quantum mechanics allows us to achieve a resolution of this difficult epistemological issue.

40. Of Single Photons and Double Slits.

Perhaps the greatest justification for the maintenance of the path-integral interpretation of quantum mechanics is supplied by the single particle variant of Thomas Young's classic double-slit experiment. The paradox of the continued appearance of interference fringes (albeit of a fuzzier nature) in this version of the experiment is normally taken as confirmation that the path integral interpretation of quantum mechanics (in which the particle is interpreted as traversing every possible path simultaneously, effectively "interfering with itself" and creating interference fringes in the process) is correct.

However, this is not the most efficient possible interpretation. My own view of the phenomenon of single particle interference fringes is that they represent a hitherto un-guessed at experimental confirmation of the existence of the "virtual particles" predicted by the Dirac equation in the 1930s. The electron or photon incidentally picks out those vacuum particles that are in phase with it, thereby generating a coherent outcome on the screen. This is why the experiment is found to work only with particles that comprise elements of the quantum vacuum, notably the electron and photon. This solution also accounts for the relative weakness of this effect when compared to traditional interference patterns created in Young's version of the experiment.

This interpretation in essence rules out the one piece of empirical evidence that seemed to lend some support to Richard Feynman's metaphysical interpretation of quantum mechanics. Furthermore, it implies that the correct interpretation of the single photon/electron experiment is *the same as* the interpretation of a whole class of other phenomena whose effects are more correctly attributed to the existence of the quantum vacuum and Dirac radiation. Most notable amongst these are the Casimir force and the Lamb shift. I would also like to conjecture that it is fluctuations in the vacuum field which contribute to individual instances of atomic decay.

The existence of this class of phenomena (which I believe should also include single particle interference fringes) points not to the validity of a particular interpretation of quantum mechanics but rather provides empirical confirmation of objective predictions derived from the Dirac equation and also from the uncertainty principle.

The reason as to why these phenomena have been correctly ascribed to vacuum fluctuation effects but single-particle interference fringes have not is because these phenomena have been discovered very much in the context of quantum mechanics and the Dirac equation. Thus, for example, the Casimir force was discovered as the result of an experiment specifically designed to test for predictions of vacuum fluctuations. But single particle interference fringes were discovered, almost accidentally, outside this interpretative framework provided by quantum mechanics. Indeed the traditional context for interpreting the results of Young's experiment has been that of illuminating the true structure of electro-magnetic phenomena. Thus it escaped a correct interpretation in quantum mechanical terms and was subsequently co-opted in support of Feynman's metaphysical interpretation of quantum mechanics instead. Nevertheless, if we were to conceive of an experiment designed to test the reality of the quantum vacuum none better than the single particle double-slit experiment could be imagined.

41. My Brane Hurts.

Future attention is likely to switch towards the possible significance and interpretation of M-theory and its objects. In principle however, new phenomena such as super-symmetry, extra-dimensions and the holographic principle make little difference to the overall cosmology offered here, except for a potential enrichment of its "fine texture" so to speak. Like any other phenomena in the universe they are all ultimately by-products of spontaneous symmetry breaking.

The situation at present however is akin to that experienced in the 1930s when Dirac's successful fusion of quantum mechanics with special relativity led to the prediction of anti-matter, which theoretical prediction was soon confirmed experimentally. The contemporary attempts to complete this task by fusing quantum mechanics and general relativity has led to the prediction of still more outlandish phenomena, phenomena required in order to escape from the infinities, negative probabilities and other mathematical irregularities that currently plague the standard model. Discrete waves and gravitons, it seems, require an additional seven dimensions in order to execute the full range of vibrations required to satisfy their equations. Just as the addition of one more dimension to Einstein's theory of General Relativity was found (by Kaluza and Klein in the 1930s) to yield a unification of classical mechanics and electrodynamics so these additional *seven* dimensions appear to be required in order to explain all species of quantum interaction whatsoever. A theory with fewer dimensions is, at present, just too simple to do the job.

This however leads to a new order of mathematical complexity never before encountered by physicists, combined with a lack of adequate empirical data due to the enormous energies required to explore and test the full range of the new theories. Taken together these difficulties could well prove an insuperable barrier to significant future progress. But this is certainly *not* to say that the general drift of the new "new physics" is wrong since it almost certainly isn't. It may simply be that our empirical knowledge is forever condemned to incompleteness in what amounts to a cosmic equivalent of Gödel's famous theorem.

Another epistemological problem derives from the fact that M-theory seems to amount to a complete set of geometries for all possible universes. In this respect it is similar to Hawking et al's idea of the wave-function of the universe, only far more potent. It is in effect a tool-box for the construction of all possible universes rather than a detailed map of the properties of our own particular one. Thus there is the seemingly insuperable problem concerning which aspects of the theory apply to our particular universe and which do not. The range of possible parameters to select from, even allowing for the maximum amount of elimination, would appear to be limitless. It is therefore, in effect, *too rich* a theory to be fully determined.

Previous paradigms for our universe supplied by Newton and Einstein had been too specific to be all inclusive, *M-theory* (which incorporates *String Theory* as a special case), by contrast, may be too general to be useful. Just as Einstein's *General Theory of Relativity* is of less practical use than Newtonian mechanics, *inspite of possessing greater generality*, so *M-Theory* is likely to be of still less practical utility, inspite of being of still greater generality. What is more, the link between generality and usefulness (a link missed by Popper) is carried over to *testability* as well. That is to say, the more fully inclusive a theory is, the harder it becomes to put it to the test. The elaborate lengths gone to by Eddington to test Einstein's theory in 1919 are a good example of this phenomenon. The anxiety felt by Popper throughout the *Logik* as to whether the *Quantum mechanics* is even falsifiable (a problem he never truly resolves) is yet further evidence. The fact that *M-Theory* is of such unique *generality* indicates that both its testability *and* its practical usefulness are profoundly open to question.

42. M-Theory and Empiricity.

Of course, throwing into question the usefulness and, in particular the *testability* of *M-Theory,* in turn raises the question of its empiricity. Popper has infact listed three criteria for a good empirical theory;

> "The new theory should proceed from some *simple, new and powerful, unifying idea* about some connection or relation between hitherto unconnected things (such as planets and apples) or facts (such as inertial and gravitational mass) or new "theoretical entities" (such as fields and particles)… For secondly we require that the new theory should be *independently testable*. That is to say, apart from explaining all the *explicanda* which the new theory was designed to explain, it must have new and testable consequences (preferably consequences of a *new kind*); it must lead to the prediction of phenomena which have not so far been observed… Yet I believe that there must be a third requirement for a good theory. It is this. We require that the theory should pass some new and severe tests."[43]

In principle *M-Theory* conforms to both of Popper's first two requirements for a good scientific theory. It represents a powerful unifying idea that connects hitherto unsuccessfully connected things; notably quantum theory and gravity. Also it predicts new phenomena (notably super-symmetry and additional dimensions), thereby fulfilling Popper's second major requirement. The obvious and common criticism of *M- Theory*, that there are not any obvious tests of the new theory, does not alter its *in principle* empirical character, since the empirical character of phenomena is independent of human ability to observe them, be it directly or indirectly. This view appears to contradict Popper's third and final criterion of a good theory however.

The problem in this case appears to be with the criterion however, not with *M-Theory*. A theory may after all be "good", i.e. empirical *and* capable of passing tests, even independently of ever having passed any tests, contrary to Popper's third criterion. The prime historical example of this is the atomic theory which was (correctly) maintained as a "good" empirical theory, independently of tests or testability, for more than two millennia. The theory was legitimately resisted over this period however, but it was always a good empirical theory. All that we can say of a theory which is not yet testable, but which is *in principle* testable, is that its status as a "good theory" is still undecidable.

Of *M-Theory* it seems fair to say that, as mentioned in the preceding section, it is a victim of its own generality. That is to say, the more general a physical theory becomes the harder it is to subject it to new empirical tests that seek to differentiate it from earlier, less general theories. In other words, as time goes by, new theory becomes progressively *harder* to verify, requiring more and more elaborate methods and equipment and overall power in order to achieve differentiation. The course of science from Archimedes through to *M-Theory* demonstrates this general rule admirably and in countless different ways, indeed, an entire research program could be devoted to its illustration, the point being that the problems faced by *M-Theory* are precisely of this kind.

At present the predictions of *M- Theory* are no more testable than those of Atomic theory were in the era of Democritus and Archimedes. The fact that they may never be *practically* testable does not alter the empirical character of this theory, notwithstanding its mathematical origins. The original Atomic theory was after all probably posited to solve a problem, not of physics, but of *logic* (to wit Zeno's paradoxes. See sections 21, 22 and 23 for a fuller discussion of these issues).

---

[43] Karl Popper, *Conjectures and Refutations,* (1963.) *Truth, Rationality and the Growth of Scientific Knowledge,* Section XVIII, p 241. Routledge and Kegan Paul. London.

This leads us to an important distinction between *Practical Testability* and *In Principle Testability*. Any good theory must conform to the latter without necessarily (at a given point in time) complying with the former. Such, we may say, is the current condition of *M-Theory*.[44]

It is often only with *hindsight* therefore (if ever) that we can be sure that a theory is metaphysical. This suggests that a possible function of metaphysics is to act as a first approximation to physical theory, with false or metaphysical elements gradually being winnowed out as a theory develops over time. Thomas Kuhn's work in particular has revealed something of this process in operation at the origin of virtually every empirical tradition in the sciences and both physics and chemistry eventually evolved out of a morass of metaphysical and quasi-empirical speculations.

All of which leads us to posit not merely limitations to Popper's otherwise groundbreaking principle of demarcation, but also the possible *undecidability of empiricity* in some cases, of which *M-Theory* might prove to be the best example.

---

[44] The principle also seems to imply that, the more testable a theory is, the more useful it tends to be. Newton's theory was eminently testable and eminently useful, Einstein's more General theory was somewhat less so. M-Theory, being practically untestable is also, by no coincidence, practically useless as well. Hence the positing of this principle linking a theory's testability to its usefulness.

43. The Problem of Decidability.

Popper, as mentioned in section forty two, was acutely aware of the difficulty of locating criteria for the possible falsification of scientific theories possessing a high degree of inclusivity and, in particular, the acute difficulty involved in establishing criteria for the falsification of general theories such as quantum mechanics whose basic statements or "protocol sentences" (i.e. predictions) are expressed entirely in the form of probabilities. As he dryly observes in the *Logik der Forschung* (hereafter referred to as the *Logik*);

> "Probability hypotheses do not rule out anything observable."[45]

And again;

> "For although probability statements play such a vital role in empirical science they turn out to be in principle *impervious to strict falsification*."[46]

Popper's implicitly heuristic solution to this problem (falsification based on some "pragmatic" criterion of relative frequency) is not a solution at all (it reminds us of latter day *Reliablism*) but an admittance of failure and a further entrenchment of the problem. Probability statements simply cannot be falsified, irrespective of the frequency with which they may be contradicted.

The correct solution to the problem of decidability is that a probability based theory (notably quantum mechanics) will, if it is genuinely empirical, be found to possess a corollary (or corollaries) which *will* be formulated so as to be falsifiable. Falsification of these corollaries is what entails the objective falsification of the entire probability based theory. Thus, probability *theories* can be falsified even though individual probability *statements* cannot. Until the corollary is falsified, therefore, the probability statements have to be assumed to be valid, irrespective of *any* degree of deviation from a standard mean. In the case of quantum mechanics (which itself constitutes the hidden foundation of nineteenth century statistical mechanics) this corollary is the uncertainty principle.

Ironically, Popper notes this very case[47] only to dismiss it. And yet the serendipitous existence of this corollary in the midst of a seemingly unfalsifiable empirical theory is an astonishing validation of Popper's overall thesis in the *Logik*.

The reason for Popper's rejection of this simple and clear cut solution was due to his personal antipathy towards the indeterministic implications of the uncertainty principle. And yet this is *inspite of* the indeterministic implications of his very own principle of falsification for the *entirety* of our scientific knowledge (quantum and classical alike), implications he never fully faced up to in his own life-time (Popper remained a staunch scientific realist throughout his life). It is a matter of yet more irony therefore that the indeterministic principle of falsifiability should have been saved from oblivion (i.e. from meta-falsification) by the existence of the indeterministic principle of uncertainty. It seems to point to a profound link between the *results* of science and its underlying *logic* of induction (as clarified by Popper), both of which are founded, (presumably for common *ontological* reasons), in *indeterminacy*, that is, in three-valued logic. Furthermore we may reasonably assume that the indeterminacy that we have discerned at the quantum origin (see section one) and which is undeniably ontological in nature, contributes a likely (ontological) explanation and indeed source for these apparent correspondences.

---

[45] Popper, "The Logic of Scientific Discovery". 1959. Routledge. Section 65. P181.
[46] Ibid, P133. (Popper's emphasis).
[47] Ibid, section 75, P218.

The fact that the principle of falsification (and demarcation) is *itself* subject to a form of falsification (i.e. in the event of quantum theory being unfalsifiable) implies that it has a self-referring as well as an empirical quality to it. In subsequent sections I shall amplify these thoughts concerning the *constructive* or "non-classical" foundation of mathematics and analytical logic in general. But the apparent self referentiality of the logic of induction itself is a powerful part of this overall argument.

Popper was unable to deduce from his misplaced distrust of the uncertainty principle an equivalent mistrust of quantum mechanics per-se. Had he done so the problem of decidability would not have arisen since the issue of the correct criterion for the falsification of an apriori incorrect or heuristic theory is all but irrelevant.

Happily though the indeterministic interpretation of the sciences is *also* saved from the status of metaphysics by the existence of the uncertainty principle. Determinism, as the fundamental interpretation of the sciences is falsified by quantum mechanics, but indeterminism is *itself* falsifiable should Heisenberg's uncertainty relations ever be violated, which, presumably, (given the generality of the theory) they never will be. This strongly implies that *interpretations* of science, since they represent deductions *from* science itself, must be classed as empirical rather than metaphysical in nature – a minor but significant point.

As a result of Popper's rejection of the uncertainty principle and of its correct interpretation (a stance he effectively maintained, like Einstein, throughout his life) the *Logik* spirals into a futile and interminable discussion of the possible logical foundations (which turn out to be non-existent of course) for theories based entirely on probability statements. This error (which is partly the result of blindness concerning the usefulness of non-classical logics) unfortunately weakens what is otherwise the single most important work of philosophy in the last two hundred years.

44. The Logic of Nature?

Induction is the primary logical procedure of discovery and the growth of knowledge, analytical logic or *deduction* is complementary to it.[48] It is the major hypothesis of much of this work (see Part two) that natural selection itself leads to the evolution of species by, in essence, *mimicking* (in a mechanical fashion) inductive logic, a *synthetical* logic ultimately based on constructive or "tri-valent" foundations. According to this model, to be analyzed in more detail in part two, each individual or species represents the equivalent of a *conjecture* (concerning the requirements for survival of that individual or species) which is then subject to *refutation* by means of natural selection. Furthermore, it is a corollary to this hypothesis that *all* forms of *cultural* evolution whatsoever are the result of humans mimicking (through their ideas and hypotheses) the above described trivalent logic of natural selection.

As just mentioned inductive and deductive logic are in truth complementary, and deductive (i.e. *analytical*) logic serves, in effect, as the foundation of inductive (i.e. *synthetical*) logic. Put formally (and this follows from Glivenko's theorem) deductive logic turns out to be a *special case* of inductive logic.

In practice however, crucial formalisms of science are *inductive* in origin, but important *corollaries* of these formalisms may be arrived at *deductively* (i.e. by means of internal analysis) at a later time. The numerous solutions to Einstein's equations are perhaps the most famous examples of this phenomenon. Black-holes for instance, which are clearly *synthetical* and *not* mathematical phenomena of nature, were nevertheless discovered purely *deductively* and not inductively by solving the equations of General Relativity for a particular case. They were deduced, in other words, as a *corollary* of an inductive theory.[49]

This example also conveniently illustrates a *relativistic* interpretation of G.E. Moore's famous *paradox of analysis*.[50] Analysis never reveals new information in an absolute sense (in other words, the paradox is veridical). But this does *not mean* that new information is unavailable from a *relative* point of view (i.e. from the point of view of the analyst). In this example the various objects which are latent in Einstein's *unsolved* equations are *not* new since they are already implicitly present in the equations irrespective of whether these are ever solved or not. The information which is brought to light by solving the equations is therefore not new from an *absolute* perspective. However, from the point of view of the persons *conducting* analysis (i.e. solving the equations for the first time in this example) new aspects of the object *are* being revealed or brought to light. Thus analysis remains a valuable activity inspite of the absolute truth of Moore's paradox. (In effect this new interpretation of the paradox reveals what might be dubbed the *relativity of analysis).*

Incidentally, the use of the term *analysis* in a *psycho-analytic* context *is* suspect since, in transforming its subject psycho-analysis does, demonstrably, produce new information. Given there are *inputs* into psychoanalysis and not merely acts of abstract, clinical analysis it would seem more accurate to characterize the psycho-analytic process as *synthetical* as well as analytical. At any rate this would at least explain why psycho-analysis seems to be as much an art as it is a science.

We may thus suggest that *induction* (rather than, for example, Hegel's "triad") represents the true underlying logic of nature and of evolution (including the evolution of culture) itself. But this will be demonstrated in far greater detail in the second part of this work.

Hegel's description of logic as laid out in his influential work "Logic" in reality represents the *inverse* of a somewhat arcane version of what has subsequently become known as "logical decomposition". Just as

---

[48] Of course we must remember that induction and deduction are *logical procedures* and are not themselves logic. They, as it were, have their *basis* in a common (non-classical) logic.
[49] Black-holes were also of course deduced from Newtonian mechanics as well, but the formal deduction from general relativity by Schwarzschild is far more precise and therefore compelling.
[50] This paradox asserts that since the elements of analysis are already contained in what is analyzed then no new information can ever be obtained through analysis.

a given complex may be *decomposed* into its parts and then again into its sub-parts, so, conversely, it may be *recomposed* and made whole again by a reverse procedure of logical synthesis. It is this *reverse procedure* of logical synthesis which Hegel's "Logic" gives us an arcane account of.[51]

---

[51] Of course, in fairness to Hegel we should note that Hegel was writing prior to the massive advances in logical analysis inaugurated by Frege. In many ways it was Hegel's reputation which suffered most from the emergence of analytical philosophy of which he may, in some respects, even be considered a precursor.

45. The Foundations of Logic.

Undecidability in mathematics represents the analytical equivalent to indeterminacy in physics. Just as Heisenberg's uncertainty relations represent the most fundamental result in the sciences (from an epistemological point of view) so do Gödel's incompleteness theorems (also two in number) represent the most fundamental result in the history of mathematics.

The first of these theorems states that any sufficiently strong formal theory (capable of describing the natural numbers and the rules of integer arithmetic) must give rise to statements whose truth value is undecidable.

The second theorem states that *if* the true statements of the formal system *are* all derivable from that system in a decidable way then it follows (given the first theorem) that the system *itself* must be internally inconsistent.

Additional axioms must therefore be adduced in order to ensure the completeness and consistency of the formal system, but these new axioms will in their turn generate new indeterminacy or new inconsistency, thereby triggering an infinite regress in the deterministic foundations of mathematics and rendering Hilbert's formalist project (the axiomatisation of all mathematics together with a finite proof of these axioms) unachievable. As Gödel himself expresses it;

> "Even if we restrict ourselves to the theory of natural numbers, it is impossible to find a system of axioms and formal rules from which, for every number-theoretic proposition A, either A or ~A would always be derivable. And furthermore, for reasonably comprehensive axioms of mathematics, it is impossible to carry out a proof of consistency merely by reflecting on the concrete combinations of symbols, without introducing more abstract elements." [52]

This in effect places indeterminacy at the foundations of mathematics, a result extended to classical first order logic by the American logician Alonzo Church a few years later.

These three results, of Gödel, Church and Heisenberg, along with Popper's work on the logical basis of induction are of definitive significance for epistemology, effectively rendering all prior epistemology redundant except as (at best) a primitive approximation to these objective, *formal* results.

This is made clear by a fifth major result – Tarski's *undecidability of truth theorem* – which extends Gödel and Church's results to *natural* languages as well, effectively implying the incompleteness or inconsistency of all classical philosophical systems.

These results therefore prove that the foundations of analytical and synthetical philosophy and reason must be non-classical in nature, in effect signaling a year zero for epistemology as well as for mathematics.

---

[52] (Lecture. 1961), "The modern development of the foundations of mathematics in the light of philosophy", *Kurt Gödel, Collected Works*, Volume III (1961) publ. Oxford University Press, 1981.

46. The Quine-White Uncertainty Principle.

Further support for the preceding arguments comes from W.V.O Quine[53]. In attacking the objectivity of the classical "synthetic-analytic" distinction vis-à-vis its applicability to the analysis of natural language in terms of predicate logic Quine effectively highlights the incommensurability of pure analytical logic and natural language, something not adequately identified by the analytical tradition prior to Quine's essay. These points were further amplified in Quine's Magnum Opus a decade later, particularly with respect to the concept of the *indeterminacy of translation* and also the so called *inscrutability of reference*.[54]

Quine ascribes this poor fit to the uncertain or *undecidable* definition of words in natural languages, a paradoxically *functional* attribute also identified by Wittgenstein.[55]

My own view is that the difficulty in utilizing these concepts un-problematically in what amounts to a post classical era stems from the fact that natural languages employ terms whose definition is largely "synthetic" in nature, as, for example, when we define a "bachelor" as an "unmarried man". Consequently, the attempt to apply formal *analytical* logic to what are in reality *synthetical* concepts leads to the problems of synonymy highlighted by Quine. In effect, the "analytical" statements in natural languages conceal a definition of terms which is "synthetical" and so the two concepts (analytical and synthetical) cannot be objectively separated in propositional logic as was hitherto assumed, a point already implicit, incidentally, in Kant's invention of the category "synthetic apriori", a category which stands at the very center of his system.

The effect is equivalent to an "uncertainty principle" residing at the heart of propositional logic. Terms cannot be hermetically defined (in accordance with the dictates of classical logic) without the possibility of *some* slippage, or *indeterminacy* of meaning. Since meanings cannot be pinned down precisely, due to the very nature of natural language synonymy cannot be taken for granted and so analyticity and determinability of translation are alike compromised. This reading was essentially reiterated two decades later by Derrida. In effect, the conclusions drawn by Quine and Derrida represent the working out of some of the logical consequences of Tarski's undecidability of truth theorem. An additional implication of this it seems to me, is that propositional logic too should be considered poly-valent in its logical foundations.

Since it is possible for the statements of natural language to closely mimic those of analytical logic this problem had not been noticed and yet Quine is correct in identifying it as of central importance since it does unquestionably explode the objectivity of the cleavage between analytical and synthetical statements in natural languages. It is therefore inadequate to irritably dismiss the effect as "marginal" since the implications for propositional logic are so revolutionary. Similarly one misses the point if, as Wilfred Sellars suggests[56] one responds to it by merely "assuming" a definition of terms which is uniform and conformal. This amounts to sweeping the difficulty under the carpet and ignores the basic point being made which is that natural languages do not function in so neat and classical a fashion. Either solution would therefore be akin to dismissing Heisenberg's uncertainty principle as a "marginal effect" that can in any event be ironed out at the classical limit.

---

[53] Wilard Van Orman Quine. "*Two Dogmas of Empiricism*". Published in "The Philosophical Review" 60. 20-43. 1951. In the interests of fairness and accuracy it should be pointed out that Quine ascribes the original idea to Morton White: "*The Analytic and the Synthetic: an untenable Dualism.*" John Dewey: Philosopher of Science and Freedom." New York 1950. P324. It may even be that the late and highly idiosyncratic work of John Dewey on logic provides the origin of Quine and White's groundbreaking insight. However, it should be pointed out that the insight, at least in prototypical form, has its ultimate modern origin in Kant's superficially oxymoronic category of "synthetic apriori" which constitutes the basis of his synthesis of classical Rationalism and Empiricism.

[54] Quine, *word and object,* 1960. MIT press, Harvard.

[55] L. Wittgenstein. "*Philosophical Investigations.*" 1951.

[56] W. Sellars. "*Is Synthetic Apriori Knowledge Possible*". Philosophy of Science 20 (1953): 121-38.

It is however legitimate to treat the distinction as valid, as it were, *at the classical limit*, or, to put it more parochially, it is valid *pragmatically,* for example, in every day speech. Objections to Quine's insight generally come from those who can see that the distinction is valid at this (essentially *classical)* level of discourse. The point Quine is simply making is that reliance on the absolute non-analyticity of the synthetic on a *theoretical level* leads to a contradiction. In essence Quine's target is Logical Positivism and in particular the work of Carnap, the essence of their system being an assertion of the complete severability of the analytic, which is *apriori*, and the synthetic, which is *aposteriori.*

The challenge of logical positivism is to Kant's system which latter system also implies that this distinction is not valid. Quine essentially sides with Kant. My own view of the ultimate significance of this, Quine's most significant contribution, is that it effectively *takes the analytical tradition back to the point at which Kant had left it*, in essence, as if logical empiricism in general had never happened.

What is needed therefore, and what this work sets out to do, is to find a *new* basis for the logicist program (whose value lies in its precision and clarity) a basis that does *not* rely on the harsh and unsustainable distinctions of logical positivism or the imprecise metaphysics of Kant. It is my contention that precisely this is supplied by recourse to *constructive* or *non- classical* logic, which retains the most important results of classical logicism, empiricism and formalism as a *special case* of a *broader* logical scheme, a scheme that accepts the inescapable and explosive category of undecidability where it arises. Indeed, this is *not* my contention; it is simply *how things are.*

Indeterminacy was thus responsible for the collapse of the first (logicist) phase of analytical philosophy, and this collapse was precipitated *primarily* by the discoveries in logic of Gödel, Tarski and Church and *secondarily* by linguistic analyses produced by Wittgenstein, Quine and many others. Nevertheless it seems to me that the limits of logic (which are ipso-facto the limits of human reason) are traced more precisely *by* logic than by either linguistic analysis or by Kant's seminal critique. As such, it additionally seems to me that these latter two can be superceded, in principle, by a *rebirth* of analytical philosophy on foundations of non-classical logic rather than on the shifting sands of linguistic analysis or the tortured metaphysics of Kant (which are the only other realistic alternatives for an epistemological basis). However, logical empiricism retains what is insightful in Kantianism, clarifying, as it does, the role of the apriori. In its updated form, rooted in non-classical logic and Popper's implicitly trivalent analysis, logical empiricism can also side step the irrelevant issue of sense datum theory since it makes no apriori assumption as to the nature or even the reality of the given (indeed the implication is that the given is inherently indeterminate in nature, a fact which is in turn responsible for epistemological indeterminacy).

47. The Rebirth of Analytical Philosophy.

The fundamental problem with propositional logic is therefore the assumption of fixed meanings for words and sentences, whereas, for functional reasons, a certain degree of **indeterminacy** is "built in" to syntax and semantics.[57] This insight concerning the intimate relationship between language (in both its syntactical and its semantical aspects) and the crucial empirical category "indeterminacy" (effectively transforming "meaning" itself into a contingent or relative phenomenon) paved the way for more extreme forms of analytical philosophy, most notably Jacques Derrida's famous "Deconstructionism"[58].
The superiority of the analytical over the continental tradition in these matters derives *entirely* from the fact that the former's arguments concerning indeterminacy ultimately stem from irrefutable *logical* analyses and empirical results (outlined in the preceding sections) rather than from the inspired (and hence disputable) intuitions (of linguistic and so called continental philosophy). Thus we are able to have more confidence in them even though they may ultimately amount to the same thing.

Although Quine claims to have identified two dogmas associated with empiricism, neither are, strictly speaking, dogmas of a more *sophisticated* species of empiricism of the sort described by Popper in the "Logik Der Forschung". Bivalence is not assumed by Popper's principle and neither, therefore, is reduction to a determinate "*given*", which is a dogma of classical empiricism rightly identified and criticized by Quine and Sellars alike. Thus the assumption that the logic of science (and indeed logic per-se) is necessarily at odds with the concept of indeterminacy is an erroneous assumption. However, this assumption has continued to hold sway in analytical philosophy since Quine's seminal attack on logical positivism.
Quine's positing of *Confirmational Holism* and *Replacement Naturalism* represents, I believe, an implicit acknowledgement of the need for *some* form of epistemology to underpin and justify analytical methodology. After all, logic is *not* linguistic and therefore language *cannot* be called the only object of philosophy.
Of course it *is* possible to maintain that logic and epistemology are not a part of philosophy, or that the concerns of epistemology stem from syntactical mistakes (as appears to be the position of Carnap and Wittgenstein) but this is not, I think, a sustainable position. Philosophy, in other words, must be concerned with *epistemological content* (i.e. the analysis of logical foundations) as well as with linguistic analysis and not just serve (vital though this function is) as the self-appointed gate-keeper of well formed sentences.
The importance of this latter function (what amounts to a *principle of clarity*) is indeed enormous however. The consequence of its absence can clearly be seen in most continental (i.e. what might be called "synthetical") philosophy since Kant. However, this on its own (as non-analytical philosophers

---

[57] Indeterminacy manifests itself in language not merely through the semantics of individual words, but through the rules and applications of grammar as well. Indeed, part of the problem with semantical theories such as Tarski's is the excessive weight they place upon the distinction between the syntactical and the semantical. It is however the *flexibility* inherent in linguistic indeterminacy which makes communication possible at all. Therefore the opposition between meaning and indeterminacy is a false one, as I have endeavored to make clear.
Consequently, indeterminacy *and* inter-subjectivity are *always* present in all public utterances. This "back-strung connection" between indeterminacy and inter-subjectivity should not surprise us however since (as Wittgenstein in particular has demonstrated) inter-subjectivity is *made possible* by the existence of indeterminacy in language use. Conversely, as our analysis of the interpretation of physics has shown (see section twenty nine on indeterministic order) indeterminacy is *always* tempered by apriori principles of symmetry and order, principles that are inescapably implied by indeterminacy. Thus the absolute distinction between indeterminacy and inter-subjectivity (or "objectivity" in the language of Kant) is unsustainable; each implies and requires the other.

[58] Consider, after all, Derrida's most famous work "De la Grammatologie" which resembles a parodic satire of Analytical philosophy. It is however far more conscious of the issue of indeterminacy than most "traditional" linguistic philosophy.

have intuited) is not sufficient justification to erect what might be called the "principle of *linguistic* clarity" as the *whole* of philosophy. The principle of clarity (whose field of application is not restricted to linguistic analysis alone) should rather be seen as *part* of the overall scheme, the informal structure of principles, of rational philosophy itself. Perhaps the ultimate *logical* justification for postulating the principle of clarity however (for those who are rightly suspicious of unjustified postulations, be they never so reasonable) is that it seems to represent a logical corollary to Ockham's famous razor[59]. The justification for Ockham's razor is apriori since it represents a logical corollary to what I call the second engineering function of the universe, which is itself apriori.

Francis Bacon, the progenitor of modern inductivism, like philosophers today (including Quine), was unaware that the relative richness of induction over deduction stems *directly* from its richer base in trivalent logic, compared to the *bivalent* base of classical logic.

The only reference to non-classical logic I have managed to track down in Quine's writing for example is primarily (though not entirely) negative in tenor and shows a profound under-appreciation of its possibilities;

> "Even such truth value gaps can be admitted and coped with, perhaps best by something like a logic of three values. But they remain an irksome complication, as complications are that promise no gain in understanding."[60]

We may observe two things about non-classical logic at this juncture. Firstly, these logics, which may operate anything from zero to an infinite number of truth values, are an unavoidable consequence of the discrediting of the principle of bivalence. This principle is *objectively unsustainable* since there are no reasons, other than prejudice or wishful thinking, to suppose that empirical and analytical statements *must* be either true or false. Indeed, we know of many examples, such as the continuum hypothesis, where this is definitely not the case, thus *objectively* overturning the classical principle. Following on from this we may observe that the objective and now indisputable discrediting of the principle of bivalence means that the whole project of *classical* logicism (from the time of Frege) was all along conducted on a fundamentally *false* basis. No wonder therefore that it abjectly failed. The fault does not lie with the *concept* of logicism however, which, in my opinion, continues to hold good.

Secondly it is by no means the case, as Quine implies, that non-classical logic is merely an irksome species of pedantry of little practical use, as it were, an irritating oddity to be duly noted prior to a return to the "more important" study of classical propositional logic. The richness of the logic of induction, which is definitely *not* Aristotelian, refutes this suggestion. It also points to a fatal limitation at the heart of elementary logic which, more than Quine's own analysis in *Two Dogmas* is ultimately responsible for the failure of logical positivism and of Frege's (implicitly bivalent) form of logicism. Furthermore, it indicates a possible root to the *reopening* of the apparently defeated logicist program on a viable basis, of

---

[59] Ockham's "principle of economy" is of supreme importance in leading to the obviation of paradoxes such as the Quine-Duhem paradox; a paradox which asserts that any empirical theory can be perpetually saved from falsification simply by adjusting it superficially. Although this is true we need not fear it because Ockham's razor implies that the emergence of simpler and more inclusive theories will tend to render such stratagems (which kept the Ptolemaic system going for centuries) ultimately futile. A simpler theory will always eventually be preferred over one which employs the elaborate stratagems alluded to by Quine and Duhem.

[60] W.V.O. Quine, *Word and Object*, MIT press Harvard, 1960, p177. It should however be pointed out that the exponential growth of work in the field of first order logic over the past one hundred years, including that of Quine himself, though far from being pointless, is of surprisingly little use. And its use is limited, in part at least, by a general reluctance to engage with the new polyvalent logic of Constructivism. Modern logic remains quintessentially *Aristotelian* in its assumptions, which entirely accounts for the frequent characterization of it as dry, infertile and *neo-Scholastic* in nature. Its faults and limitations are indeed precisely those of Aristotle's *Organon* as noted by Sir Francis Bacon. As Bacon correctly recommended (trivalent) inductive logic as the cure for the problem so we seek to propose constructivism as the cure for the problems faced by modern logicism after Frege.

which more will follow in later sections of this work. This, which I call *neo-logicism,* is the *only* viable alternative to Quine's unsustainable suggestion of "natural" epistemology.

In the absence of a *synthetical* dimension (which is supplied to empiricism by means of the logic of *induction*) a philosophy cannot consider itself truly complete. *Empiricism,* for example, is a powerful example of the equipoise of the analytical and the synthetical, as this work hopefully demonstrates. It is the great advantage of non-classical over classical logic that it allows for a fully *synthetical* as well as a fully analytical philosophy. Ultimately however it reveals these distinctions (so ingrained in modern philosophy) to be illusory.

The effect of the demise of the postulated cleavage between the categories "synthetical" and "analytical" is the apotheosis of indeterminacy or undecidability (which we have already encountered with reference to the logic of quantum mechanics and mathematical incompleteness), a fact which is indeed fatal to the pretensions of classical empiricism, based as it is on the bivalent assumption of the principle of verification.

But it is clearly *not* fatal to a Popperian *neo-empiricism* based as *it* is on the *three* valued logic that implicitly underpins Popper's new principle of falsifiability. Popper is in effect the saviour of the logic of empiricism which Carnap had tried and failed to save on the basis of classical logic and which Quine has also failed to find an adequate (non-metaphysical) justification for. Indeed, as mentioned earlier, all these failures have merely taken philosophy back to the Kantian synthesis. Only Popper's discovery takes us somewhat beyond this point.

As such, Popper's analysis in the *Logik der Forschung* is one of the key results of analytical philosophy. Indeed, since Popper and Gödel, the epistemological foundations of analytical philosophy, like those of empiricism itself, have quietly shifted from the assumption of *classical* logic to that of multi-valent logic, which is *also* the new logical basis of *induction* itself after Popper.

Linguistic philosophy, with its happy embrace of the indeterminacies of natural language use, can be reinterpreted therefore as the *informal* (in effect *complementary*) recognition of this new epistemological basis, a basis which can, nevertheless, still be characterized *entirely* in terms of (polyvalent) logic. The goal of *Principia Logica* is thus to make the already *implicit* acceptance of non-classical logic (for example in linguistic analysis) *explicit*. It will then be seen that the true foundation of philosophy and epistemology is not primarily *linguistics* but rather (non-classical) logic.

48. The Incompleteness of Empiricism.

An additional benefit, mentioned earlier, of what I call "neo-empiricism" (i.e. empiricism in the light of Popper's logical analysis), is that it renders Quine's *metaphysical* solution to the problem of induction (i.e. so called "conformational holism") unnecessary. Since, thanks to Popper (who admittedly did not see it in *quite* these terms) indeterminacy is fully incorporated into the logic of induction it therefore follows that the supposed need for the salvation of the principle of verification through the doctrine of "conformational holism") is eliminated. After Popper, the *problem of induction*, which Quine's holistic theory sought to address, simply does not exist anymore, since the correct logical foundations of empiricism have already been established by Popper, be it never so unwittingly.

Popper has in any case detected logical flaws in the doctrine of conformational holism which are fatal to it. Popper's decisive, yet largely unappreciated, demolition of *Confirmational holism* is contained in the following passage;

> "Now let us say that we have an axiomatized theoretical system, for example of physics, which allows us to predict that certain things do not happen, and that we discover a counter-example. There is no reason whatever why this counter-example may not be found to satisfy most of our axioms or even all of our axioms except one whose independence would thus be established. This shows that the holistic dogma of the "global" character of all tests or counter-examples is untenable. And it explains why, even without axiomatizing our physical theory, we may well have an inkling of what went wrong with our system.
>
> This, incidentally, speaks in favour of operating, in physics, with highly analyzed theoretical systems – that is, with systems which, even though they may fuse all the hypotheses into one, allow us to separate various groups of hypotheses, each of which may become an object of refutation by counter-examples. An excellent recent example is the rejection, in atomic theory, of the law of parity; another is the rejection of the law of commutation for conjugate variables, prior to their interpretation as matrices, and to the statistical interpretation of these matrices."[61]

It is also apparent that the rules of grammar, like the axioms of mathematics, are capable of exhibiting a high degree of autonomy from one another. We may understand some elements of grammar without necessarily understanding others (otherwise grammatical mistakes would never arise). As such the metaphysical doctrine of *semantic holism* is also untenable, especially as the "whole of language" is no more a clearly *determinate* thing than is the "whole of science".

Granted that the systems permitting language *aqquisition* are innate (due to evolution), nevertheless it is probably the case that no-body understands the *whole* of any language as it is not even clear what this would mean. For this reason the popular concept of *Holism* per-se is unsustainable and should be replaced by that of *indeterminacy*, which latter is more analytically rigorous.

The ultimate failure of classical Empiricism and Rationalism alike stems from efforts to *exclude* the constructive category of undecidability. By contrast what I call *neo-empiricism* (which reinterprets the logic of induction and analysis alike in the light of *Constructive* logic) provides a rigorous formalization of this concept. Nevertheless, neo-empiricism per-se is *not* sufficient of itself to constitute the epistemological basis of the sciences.

---

[61] Karl Popper. *Truth Rationality and the Growth of Scientific Knowledge.* 1960. Printed in *Conjectures and Refutations,* Routledge and Kegan Paul press, London. 1963.

Empiricists would have it that the concept of *existence* is *aposteriori* in nature. However, because the notion of "absolute nothingness" can be demonstrated to be *logically* meaningless (a "non-sequitur" so to speak, similar in nature to Russell's set paradox) it therefore follows that "*existence*" per-se (in its most general and to some extent *empty* sense) *is* an *apriori* concept, a fact alluded to by Kant's decompositional analysis (vis-à-vis the so called "categories" and the "forms of intuition"). And this flatly contradicts the assumptions of classical empiricism.

At any rate, an updated defence of apriorism, even in the context of neo-empiricism, does seem to follow from the logical and mechanical considerations discussed in the earlier sections of this work, particularly as they surround what I identify as the two engineering functions of the universe.[62] It is of course a tribute to Kant's acuity that his system foreshadows these elements of neo-logicism and neo-empiricism.

---

[62] Nature is freely *constructive* in the sense that it is *self organizing*, due to the apriori pressures exerted by the two engineering functions, which are themselves deducible apriori, as analytical truths. The validity of the constructivist interpretation of mathematics perhaps stems from these *analytical* foundations.

49. The Epistemological Basis.

Epistemology is interpretable as the logical basis upon which philosophy is conducted. This basis can only be of two types; classical (or "bivalent") and constructive (or "multivalent"). All epistemology of all eras is therefore reducible to one or other of these two logical foundations. Indeed the main thrust of analytical philosophy prior to the Second World War can be characterized as the failed attempt to permanently secure philosophical analysis on *classical* logical foundations. In retrospect the failure of this project can be attributed to the implications of Gödel and Church's theorems on the one hand and Popper's implicitly three valued analysis of induction on the other.[63] These discoveries demonstrated once and for all that epistemology can *never* be based on bivalent or Aristotelian logic as had previously been assumed by Rationalists, Empiricists and indeed Kantians alike.[64] Dim recognition of these facts triggered the transformation of Analytical philosophy from a logical to a linguistic focus along with the implicit acknowledgement of indeterminacy (and thus multi-valued logic) which this implies.

The beauty of many valued logic is that it is *not* axiomatizable and therefore is not subject to Godelian incompleteness due to its recursive or self-referring quality. It has incompleteness "built in" so to speak and so is subject only to itself in a finite fashion. *Its foundations are therefore* **tautological** *rather than axiomatic, which amounts to saying that indeterminacy is* **itself** *indeterminate and therefore logically consistent.* In short it is *apriori*.

So called *intuitionistic* (i.e. constructive) logic is thus, in essence, the true logical basis of indeterminacy and therefore forms the correct epistemological foundation of philosophy (including what I call *neo-empiricism,* by which I mean empiricism in the light of the principle of falsification). Indeed, this common epistemological foundation also points to the underlying unity between the synthetic and the analytic, between inductive and deductive reasoning, (a "back-strung connection" as Heraclitus would have said). This connection was finally proven by Glivenko.

We may therefore fairly assert that the abandonment of the attempt to express epistemological foundations in terms of pure logic was premature. Although these foundations turn out not to possess the assumed (bivalent) form, they nevertheless clearly *do* exist and it is therefore an important function of Analytical philosophy to investigate and to recognize this fact. In any case, to reject constructive logic is to reject *any* viable basis at all for epistemology, thus needlessly opening the door to absolute (as distinct from contingent) scepticism.[65]

Nevertheless, constructive logic, inspite of supplying the epistemological basis for neo-empiricism, is itself without absolute foundations. This paradoxical situation (which, as just mentioned, amounts to the near tautology that indeterminacy is itself indeterminate) subsists for *ontological* reasons and should be what we expect in view of the phenomenon of universal relativity. If this ontological explanation for what I term the *fundamental paradox of epistemology* is correct then it strongly implies the underlying *unity* of epistemology and ontology. Epistemology, according to this understanding, would be nothing more nor less than a special (albeit privileged) form of ontology. Or, to put it more bluntly, epistemology

---

[63] Officially Quine is given the credit for its demise, but Positivism had infact already collapsed more than a decade prior to the publication of *Two Dogmas*.

[64] The non-classical epistemological basis of mathematics is clarified by Gödel's incompleteness theorem, but, much as a tree is supported by its roots the epistemological foundations of mathematics are supplied by constructive logic. In essence classical logic is a special case of Intuitionistic (i.e. constructive) logic.

[65] Popper has in fact criticized the application of constructive logic to the problems of epistemology (see especially *Truth, Rationality and the Growth of Scientific Knowledge*, section 10, published in *Conjectures and Refutations*. Routledge and Kegan Paul. London. 1963.) But the arguments he adduces are unconvincing and (in his revisionistic search for the comforts of classical foundations) he appears not to notice the fact that his own non-trivial contribution to epistemology is itself implicitly based on trivalent logic.

Indeterminacy cannot be evaded in either ontology or epistemology, neither can it be coped with in terms of classical logic or in schemes (such as Tarski's "Semantical" conception of truth, or Logical Positivism) which are implicitly based on classical assumptions. Of this much we need be in no doubt.

is the way it is ("grounded" in the constructive logic of indeterminacy) for fundamentally *ontological* reasons, suggesting therefore the ultimate priority of ontology over epistemology. But of course none of these conclusions can be definitively verified, by definition. If they could be so verified (according to the scheme of classical logic) then they would ipso-facto be false. Only the recursive consistency of constructive logic, coupled with the ontological implications of quantum theory demonstrate the veracity of these things.

Further corroboration of the ultimate unity of ontology and epistemology *is* however provided by the analysis of the equally central role played by indeterminacy in quantum mechanics and also by the physical interpretation of the concept of indeterminacy discussed in section one of this work. In essence indeterminacy can be given a precise physical or mathematical definition as (quite simply) zero divided by zero. As such this fundamental epistemological category would appear to possess an ontological significance as well, pointing to what I have termed *ontological nihilism*. Indeed the *epistemological nihilism* made famous by Nietzsche and by modern philosophy in general can be interpreted as nothing more than a *special case* or logical corollary of what is a *far more* consequential state of affairs – that of onto-nihilism. And onto-nihilism is nothing more than the expression, in terms of ontology, of the empirical and analytical category of indeterminacy. After all, undecidability can easily be interpreted as indicating the absence of anything *substantial* to decide.

In essence, therefore, the problem of epistemology cannot be entirely resolved *outside* the context of ontology. The problem of modern philosophy since Descartes has largely stemmed from the attempt to resolve the (very real) problem of epistemology in splendid isolation – something which simply cannot be done.

At any rate, this fundamental unity of ontology and epistemology – joined at the hip by the all purpose category of indeterminacy – is a unique and non-trivial result of neo-empiricist philosophy.

50. Three-Valued Logic.

Truth and falsehood in three-valued logic (a form of logic that naturally arises as a consequence of *Intuitionistic* or *Constructive* assumptions)[66] are not absolutes, but represent limiting cases of indeterminism. In diagrammatic form the situation can be represented thus;

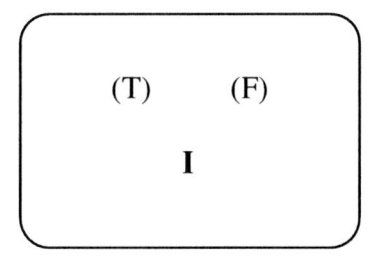

```
┌─────────────────┐
│                 │
│   (T)     (F)   │
│                 │
│        I        │
│                 │
└─────────────────┘
```

Truth (T) and falsehood (F) can, for practical purposes, be represented as clearly distinct from each other but (given the discoveries of Gödel, Church, Heisenberg, Popper and others) not *absolutely* so. This is because the logical landscape they inhabit is one of a far more general, indeterminate {I} character. This is due to the absence of the Aristotelian *law of excluded middle* in constructive logic. To use a convenient analogy *truth* and *falsehood* bear the same relationship to each other in three-valued logic that two waves in the sea do to one another, with the sea in this analogy symbolizing *indeterminacy*. Clearly waves are *part* of the sea and, likewise, truth and falsehood are ultimately *indeterminate* in nature.[67] In the case of many-valued logic exactly the same set of relationships obtain, with each value in the system being like a wave in the sea of indeterminacy. In multivalent logic there can, so to speak, be an *infinite* number of such waves;

```
┌─────────────────┐
│                 │
│  (1), (2), (3), (4), │
│   (5), (6)… etc.│
│                 │
│        I        │
│                 │
└─────────────────┘
```

It should be clear that this is a richer model of logic than that provided by the discrete, bivalent model that has held sway since the time of Aristotle. Indeed, this greater richness (which should be intuitively apparent) has been objectively demonstrated in a little known theorem by the great Russian mathematician Valery Glivenko. In this obscure but crucially important theorem Glivenko succeeds in proving that constructive (i.e. Intuitionistic) logic contains classical logic as a special case of itself – thus demonstrating the greater generality (and epistemological importance) of constructive logic.[68]

---

[66] The doctrine of Intuitionism represents an application of Kantian epistemology to the philosophy of mathematics and was made popular by the Dutch mathematician L.E.J. Brouwer at the beginning of the twentieth century.
 The doctrine, as a philosophy, is irrelevant to a technical understanding of the basis of constructive logic (which is simply the rejection of *tertium non datur*) but in essence expresses the neo-Kantian view that the positive integers and their arithmetic are (like the cognitive structure of the human mind itself) presupposed by the transcendental Kantian categories.
[67] Using the language of logic (somewhat comically) in this analogy we might say that waves are like "limiting cases" of the sea.
[68] Sadly, outside the Russian language, information on Glivenko's theorem is almost impossible to locate, which is a strange state of affairs given the manifest epistemological significance of his result, which I have sought to bring to light in this analysis. Technically speaking what Glivenko succeeds in proving is that a formula "A" is provable in the classical propositional logic if "~~ A" is provable in the constructive propositional logic. In other words, any valid classical formula

In a footnote to this discovery it may be pointed out that, on account of the work of Popper on the principle of falsifiability, the logic of induction has been shown to be three-valued. Given that the logic of deduction is classical or bivalent in nature it therefore follows that the neo-empiricist (three-valued) logic of *induction* must contain the classical (bivalent) logic of *deduction* as a *special case* of itself, effectively demonstrating the underlying unity or "back-strung connection" that exists between analysis and synthesis, i.e. between deduction and induction. The crucial ancillary role played by deduction in explicating the contents of inductive theories merely serves to corroborate the validity of this inference. This specific implication of Glivenko's general theorem is of surpassing importance for the epistemology of science, just as the theorem *itself* is for the implied unity of epistemology per-se.

It is also worth noting at this point that the most important pieces of analysis for epistemology (Heyting's Intuitionistic logic, Lukasiewicz's many valued logic, Gödel, Church, Tarski and Glivenko's theorems, together with the three principles supplied by Popper and Heisenberg) have occurred *outside* the context of analytical philosophy and philosophy per-se.[69] Other than perhaps Quine's recursive analysis of analyticity it is impossible to think of a single insight of like importance to epistemology arising out of philosophy (of any type) over the same period. It is additionally regrettable that philosophers have in general fought shy of analyzing the implication of this torrent of new and uniquely objective insights to their subject.[70] Rather than the retreat from epistemology into linguistic analysis (in light of the failure of logical positivism), philosophy *should* primarily (though not solely) have focused on explicating the full implications Intuitionistic logic and these new theorems and principles for the objective foundations of their subject. Either these formalisms have *no* implications for epistemology or else there has been a dereliction of duty on the part of most analytical philosophers still stung by their abject failure to place all knowledge (like a fly in a bottle) within the impossible confines of classical logic.

---

can be proven in the constructive propositional logic if a double negative is placed in front of it, thus demonstrating the latter to be a more general and complete form of logic.

As a result of Glivenko's theorem (and Heyting's *Intuitionist* logic) neo-empiricism can be considered to be logically complete and well founded (thereby fulfilling the seemingly long dead ambition of logical empiricism). Not, I believe, a trivial result.

[69] See for example, Jan Lukasiewicz's ground breaking work on many-valued propositional calculus contained in his *Elements of Mathematical Logic* Warsaw 1928, English translation, PWN, Warszawa, 1963.

Many-valued logic follows as a possibility from the assumptions of the Constructive (i.e. Intuitionistic) logic of Brouwer and Heyting. It is the abandonment of the law of excluded middle (this abandonment is the basis of all constructive logics) which gives rise to the possibility of many-valued logics. Out of the immolated corpse of Aristotelian logic (so to speak) arises the infinite riches of Constructive logic.

What is of interest in non-classical logic is its greater *generality*, which flows from the simple procedure of abandoning the law of exclude middle as an unjustified and unnecessary assumption. We may say that the law of excluded middle (which retains the law of contradiction as a necessary corollary) is not merely unjustified (a fact which constitutes the basis of most criticism of it) but that it is also *unnecessary* because intuitionistic logic retains classical bivalent logic as a special case.

I might add that there is clearly a pattern to all these results and analyses which collectively point to the emergence of a uniquely objective foundation for epistemology (both analytical and synthetical alike) supplied by constructive logic. This, in my view, constitutes the most important development in the history of epistemology and it is almost completely unobserved by philosophy, even Analytical philosophy.

[70] For an exception to this general trend see M. Dummett. *The Logical Basis of Metaphysics,* Harvard University Press. 1991.

51. Pyrrho and the End of Greek Philosophy.

To call these formalisms the most important development in the history of modern epistemology is not to single them out for their uniqueness of insight, but rather to assert their unique contribution in placing the previously disputable category of indeterminacy onto indisputably objective foundations. This is the true shift that has occurred (indirectly) by way of analytical philosophy.

The specifically "philosophical" understanding of the centrality and implications of indeterminacy however dates back more than two thousand years to the time of Pyrrho of Elis, a contemporary of Aristotle who is reputed to have accompanied Alexander the Great on his notorious march into Persia and India.

Pyrrho, according to Aristocles' second century C.E. account of the views of Pyrrho's pupil Timon of Phlius, is believed to have deduced indeterminacy (*Anepikrita*) as a consequence of his doctrine of *Acatalepsia* or the inherent unknowability of the true nature of things.[71] "For this reason" Aristocles asserts "neither our sensations nor our opinions tell the truth or lie."[72] Expressed in terms of modern analytical philosophy this doctrine might be restateable along these lines; "*Analytical and synthetical statements are alike indeterminate,*" which is the *objective* conclusion that contemporary philosophy has (in my opinion) been brought to by virtue of the universality of non-classical logic. What *is* unique about the present situation however is that it is now highly formalized, which was not the case in the era of the Middle Academy.

According to Pyrrho the awareness of indeterminacy (*Anepikrita*) leads to *aphasia* ("speechlessness")[73] and hence, logically, to the suspension of judgment (*epoche*) on specifically epistemological matters (though not necessarily on practical matters).[74] From this state of mind *ataraxia* ("freedom from cares") is believed to arise naturalistically, leading to the famous Pyrrhonic attitude of tranquility and imperturbability, even in the face of great tribulations.[75]

Pyrrho is said to have made "appearances" the true basis for all practical action – implying perhaps the world's first recorded instance of a purely pragmatic basis for ethics. In modern terms we might call Pyrrho, given his belief in the utility of appearances, a *phenomenalist* (as distinct from a "phenomenologist"). His view of the matter would seem to be essentially correct since *phenomena* are

---

[71] Pyrrhonism is a doctrine arrived at by means of a critique of what today would be called the "the Given". Though lost to us this critique may have been similar to Arcesilaus' attack on the Stoic doctrine of *catalepsia*, the belief that determinate epistemological knowledge can be gathered by means of direct sensory perception. In any event it is clear that the problem situation faced by the Academies and Schools in the immediate wake of Pyrrhonism was almost identical to that faced today in the era of Analytical philosophy, with Stoicism playing the role of logical positivism.

[72] Eusebius, *Prep. Ev.* 14.18.2-5, Long, A.A. and Sedley, D.N. *The Hellenistic Philosophers.* 2 volumes. New York, Cambridge University Press, 1987.

[73] This perhaps reminds us of Wittgenstein's views.

[74] Thus Aristocles; "if we are so constituted that we know nothing, then there is no need to continue enquiry into other things." Eusebius, 14.18.1-2, Long and Sedley, op cit.

[75] It is possible that the Stoic doctrine of *catalepsia* was posited in response to the Pyrrhonic doctrine of *acatalepsia*.
If this is indeed the case then it may *also* be the case that Arcesilaus fashioned his critique of Stoicist epistemology as a defense of orthodox Pyrrhonism. The view (attributable to Sextus Empiricus) that an absolute distinction can be drawn between Pyrrhonism and Academic scepticism is not really sustainable. Fortunately we have Diogenes Laertius' account that Arecesilaus; "was the originator of the Middle Academy, being the first to suspend his assertions owing to the contrarieties of arguments, and the first to argue pro and contra" (4.28-44, Long & Sedley. Op Cit). This suggests that Academic scepticism was a synthesis of Sophism (including the views and dialectical methods of Socrates) *and* Pyrrhonism (i.e. the acceptance of indeterminism together with the "suspension of judgement", i.e *epoche*). This is almost certainly the truth of the matter.
This interpretation implies that Socrates was profoundly influenced by the Sophist views of, for example, Protagoras and Gorgias, a fact which seems to have been suppressed by Plato who sought to use Socrates (the historical figure) as a mouthpiece for his increasingly elaborate metaphysical theories. What we know of Socrates independently of Plato indicates that he may have been a Sophist and a sceptic. Consequently, the switch from orthodox Platonism to Analytical Scepticism on the part of the middle academy could be interpreted as a simple reversion to Socratic Sophism.

the *direct consequence* of indeterminism. As such we can objectively say that Pyrrho's alleged *practical* commitment to "appearances" (i.e. "phenomena") as the basis for all action does not inherently contradict his *epistemological* commitment to indeterminacy.[76] This is because indeterminism does not stand in *opposition* to phenomena. Instead indeterminism is what is ultimately *responsible* for all phenomena, including the appearance of order and symmetry (being those properties which make inter-subjectivity possible)[77]. Thus, as it were, "going with the flow" of appearances (phenomena) is not inconsistent with a contingent commitment to epistemological uncertainty.

The fact that the status of indeterminacy is *itself* indeterminate is not necessarily a contradiction of indeterminacy – it is more like a tautology. The "indeterminacy of indeterminacy" does not magically render *anything else* unconditionally determinate. Neither does the "doubting of doubt itself" render anything else proven. Rather, it tends towards entrenching uncertainty still deeper. Indeterminacy is, in this purely *contingent* sense, absolute. This is also the correct technical defence of scepticism against the charge of inconsistency and self refutation. (In effect we need to distinguish between *bivalent scepticism* (which *is* inconsistent) and *polyvalent scepticism* (which is not.))

What is distinctive about Pyrrhonic scepticism is the rigorous way it calls doubt itself into question. It is not therefore *bivalent* in the way of a more traditional and outmoded conception of scepticism. It does not definitively and inconsistently assert the unknowability of truth in quite the way that earlier Greek skeptics (notably Xenophanes) had done. It is the most subtle form of scepticism of all and it exactly captures the subtlety of modern non-classical logic.

Notwithstanding the possible influence of the likes of Heraclitus, Democritus, Gorgias and others on Pyrrho's system it nevertheless seems reasonable (not least, but also *not only*, because the sources tell us so) to locate perhaps the major influence and motivating force behind the construction of Pyrrho's system (or the system traditionally attributed to Pyrrho) to India. Pyrrho's trivalent or *constructive* outlook, although not unknown in Greek philosophy at that time (see Heraclitus for example) was infact the orthodox epistemological position in Indian philosophy from that era, as is evidenced by the Upanishads (c800-650 BCE). Indeterminacy represents the logical conclusion of a rigorous method of doubt called the *Sarvapavada* or the method of absolute negation. Through its rigorous application Hindu philosophers arrived at the logical conclusion that *neither doubt nor certainty are establishable* – a position which describes the essence of Pyrrhonic (and indeed modern) epistemology.[78]

---

[76] This outlook is somewhat reminiscent of Carnap's instrumentalism, as expressed in the following passage;

"Thus it is clear that the acceptance of a linguistic framework must not be regarded as implying a metaphysical doctrine concerning the reality of the entities in question. It seems to me due to a neglect of this important distinction that some contemporary nominalists label the admission of variables of abstract types as "Platonism." This is, to say the least, an extremely misleading terminology. It leads to the absurd consequence, that the position of everybody who accepts the language of physics with its real number variables (as a language of communication, not merely as a calculus) would be called Platonistic, even if he is a strict empiricist who rejects Platonic metaphysics." Rudolf Carnap, *Meaning and Necessity: A Study in Semantics and Modal Logic,* enlarged edition (University of Chicago Press, 1956).

[77] Determinism, as we have seen, need not be rejected, but should instead be interpreted as a special instance or limiting case of *indeterminism*. In ontological terms this may contingently be interpreted as meaning that what we call *existence itself* is a limiting case of inexistence – i.e. there is no substantial or eternally subsisting being or essence, only transience and universal relativity, which express themselves in the form of *phenomena*.

[78] Flintoff (*Flintoff, E. (1980), 'Pyrrho and India', Phronesis 25: 88-108.)*.has been influential in arguing the case for the Buddhist influence on the formation of Pyrrhonism, but an Hindu source seems to me more probable. The account left by Megasthenes (who was the representative of Seleucis Nicator in the court of Chandragupta Maurya some twenty years after the death of Alexander) strongly suggests that the two major cults of Hinduism – Saivism and Vaisnavism – were *already* strongly established at the time of Alexander (see Arian, *Indica* 8.4, 11-12, which mentions Mathura – center of the cult of Krishna – as the headquarters of one of the cults and also see Strabo *Geography* Book one, 15.1.58, which mentions the Himalayas as the center of the other cult). Since the cult of Rudra Siva has always been centered in the Himalayas it is highly likely that Alexander's men would have had contact with Sadhus (*Gumnosophistai* as the Greeks called them) from this sect. Furthermore, it is the case that the Saivists have always been the primary exponents of the philosophical method known as *Sarvapavada*. Hindu Annihilationism has thus come down to us from these philosophers who are also closely associated with

The method of absolute negation (*Sarvapavada)* was also used to develop an ontological position to the effect that neither being (i.e. "permanent substance" in the Aristotelian sense) nor nothingness (in the sense of "absolute void") truly exist – an ontology (because it does not lead to reification or hypostatization) that might fairly be described as *hyper-nihilistic*.[79] It is this ontology and this epistemology (collectively known as *Annihilationism*) which effectively underlie what has since come to be thought of as Buddhist philosophy.[80]

Inspite of its generality and logical correctness Annihilationism (*Sunyavada*) has remained largely unknown in the West down to the present day, and yet, via Pyrrho, it has probably exerted a decisive influence on the course of Western philosophy, both ancient and modern. Indeed it may even be true to say that from a logical point of view and probably also from a historical point of view as well Western philosophy (like Buddhist philosophy) is little more than a special case of Indian philosophy. Not for nothing does Sextus label Pyrrhonism the most powerful system of the ancient Greco-Roman world. And yet (like Buddhism) it is little more than an off-shoot of Hindu epistemology. Furthermore, *Sarvapavada*, as a developed method and doctrine, was already established prior to the accepted origins of Greek philosophy in the seventh century BCE.[81]

Pyrrho, inspite of his relative obscurity, is thus a pivotal figure in the history of Western philosophy. He effectively signaled the end of the significant development of Classical philosophy, after which it

---

the origins and the later development of Buddhism. Thus it seems likely to me that a Saivist influence on Pyrrhonism is at least as probable as a Buddhist one.

An additional point to make is that Pyrrho also introduced the doctrine of *apatheia* or "inactivity" into Greek philosophy, arguing that it, like *ataraxia* "follows like a shadow" upon the suspension of judgment. Again this is a notion somewhat alien to Greek philosophy of the era, but not at all alien to Hindu philosophy.

At any rate it might at least be asked why, if Pyrrhonism does not originate in India, were there so many Pyrrhonists (i.e. Annihilationist Sadhus) in India *prior* to the time of Alexander and so many Pyrrhonists in Greece thereafter?

[79] This represents the most comprehensive solution to the problem of ontology on purely rational grounds. Furthermore, it is a complete and correct solution, one that still remains for us to absorb. It therefore deserves mention alongside and even ahead of Parmenides rational deduction of the block-universe and Democritus' rational deduction of atomism, both of which deductions were *also* based on logic alone. Sunyavada nihilism merits precedence due to the greater *generality* of its solution to the twin problem of ontology and epistemology. The correct (formalistic) view is therefore to treat the cosmology of Parmenides and the physics of Democritus as subordinate parts of the overall system supplied by Sunyavada nihilism, notwithstanding the absence of any historical connection (there isn't one). They are (in modern terms) the physics to *its* epistemology so to speak.

In any event this ancient achievement refutes the implication of Kant's epistemology which is that a correct ontology and cosmology cannot be supplied on the basis of pure reason alone. It can be and it has been. Empiricism has merely served to *confirm* the general correctness of the cosmology of Parmenides and Democritus and of the Sunyavada ontology. Empirical method has thus served the incidental function of "deciding between the systems (both ancient and modern)" so to speak. But these systems were hypothesized on *purely rationalistic* grounds, contrary to what is deemed possible according to Kant's *Critique of Pure Reason*.

Another example of the limitations in Kant's critique of Rationalism concerns his classic cosmological antinomy between the finite and the infinite. This antinomy is infact obviated with reference to the sphere. One might therefore say that the antinomy was solved by Riemann and by Einstein – in effect through *applied* Rationalism (which is what Empiricism truly amounts to).

[80] In the Buddhist system the search for *Truth* concludes in the recognition that *truth* is indeterminate due to *universal relativity*. This amounts to holding to a relative and not an absolute view of the nature of truth. The doctrine of universal relativity (that entities lack inherent existence due to "causes and conditions") is the distinctively original contribution of Buddhist analysis to the school of Annihilationism (*Sunyavada*). The finest expression of this overall synthesis is to be found in the philosophical works of Nagarjuna, who is the pinnacle of Indian, Buddhist and, in many ways, world philosophy. Of particular importance are the "Seventy Verses on Emptiness" and the Mulamadyamikakarrika.

[81] It is clear that Hindu logic has no truck with *Tertium non Datur*, that it is implicitly constructive or non-classical in nature. Nevertheless, the Vedantic philosophy expressed in the *Upanishads* is technically at fault with respect to the logic of Sarvapavada. This is because the Sarvapavada undermines the tenability of the doctrine of the Atman as it is dogmatically asserted in the *Upanishads*. This indeed is the basis of the critique of Vedantism effected by the Buddha. As such Buddhism rather than Vedantism must be considered the true inheritor of the tradition of the Hindu Annihilationist philosophers and of the method of negation.

The power of the Buddhist critique of Vedantism is implicitly reflected in the reform of Vedantism made by the Hindu philosopher Shankara in the ninth century. This reform is encapsulated in the concept of the *Nirguna Braman* or "attributeless spirit".

Shankara's philosophy, more than any other defense of theism, unintentionally demonstrates how the belief in a transcendent, formless, yet all powerful being is only ever one step away from complete ontological nihilism.

devolved into a set of competing but essentially unoriginal schools.[82] In much the same way, the institution of constructive logic at the heart of Analytical philosophy marks the end of the significant development of philosophy in the modern era. Of this there can be absolutely no doubt.

Pyrrhonism went into decline soon after the time of Sextus Empiricus. The epistemological dogmas of Pyrrho's great contemporary, Aristotle, better suited an age of faith than Pyrrho's *hyper-scepticism* ever could. Aristotle's more limited and, it must be admitted, *formalized*, view of logic similarly triumphed and has only recently been overturned as a direct result of the successful formalization of constructive logic.

Nevertheless, ever since a form of Pyrrhonism overtook Plato's academy, Pyrrhonism (more so than orthodox Platonism) has alternated with Aristotelianism for epistemological primacy. This alternation represents the underlying thread of continuity in Western philosophy down to the present time. In essence, the period from the Renaissance through to the demise of analytical positivism constitutes a period of slow transition from the Aristotelian to the Pyrrhonic view. It is this transition which, after the demise of classical Logicism, is now largely complete. Given the long hegemony of Aristotelianism we can but expect an *exponentially longer one* for the Pyrrhonic viewpoint since it is a more complete viewpoint that, as it were, retains Aristotle's classicism (classicism of all types) as a special case.

The beginning of this epistemic change may be said to originate with Montaigne whose influence on Descartes is undisputed. Descartes hugely influential "method of doubt" may reasonably be traced back to Montaigne (whose scepticism, in large part, derives from Sextus' "Outlines of Pyrrhonism"). Thus Descartes' inception of modern Rationalism can fairly be interpreted as an attempt (continued, according to one interpretation, by Kant) to refute Montaigne's Pyrrhonism. Kantianism, on this view, can be interpreted as the (now largely redundant) defence, not only of classical empiricism, but, more fundamentally still, of Aristotelian epistemology.

Similarly, Hume's demolition of classical empiricism may be said to spring from his adoption of Pyrrhonic scepticism, whose merits he repeatedly stresses. Indeed, Hume's brilliant adduction of the problem of induction (and hence of the limits of causality) constitutes the only truly novel argument for scepticism to appear since Sextus.

It is therefore not wholly unreasonable to regard "Outlines of Pyrrhonism" as the most influential work in the history of philosophy, notwithstanding its relative obscurity. Thus without "Outlines of Pyrrhonism" modern philosophy as we know it (Rationalism, Empiricism and Kantianism alike) simply would not exist.[83]

---

[82] Chrysippus is perhaps the one true exception to this sweeping generalization. See Lukasiewicz, Jan: *Aristotle's Syllogistic, from the Standpoint of Modern Formal Logic*. Clarendon Press. Oxford, 1951.
[83] Infact, of these three movements only empiricism originates from sources wholly uninfluenced by Pyrrhonism. Bacon's revival of classical empiricism stems from a reaction against Aristotle's *Organon*. Nevertheless, Hume's deadly attack on classical empiricism *does*, I feel, stem from his Pyrrhonism.

52. The Second Truth Theorem.

Of the various formalisms previously discussed as objectively pointing to a constructive (and hence relativistic) epistemological basis Alfred Tarski's so called *undecidability of truth* theorem is of particular importance. This is because it allows us to generalize Gödel's *second* incompleteness theorem so that it applies to the expression of truth in natural languages. In effect Tarski's theorem states that; *every universalistic language generates internal inconsistencies* (for instance, paradoxes such as the liar's and the barber's).
As Popper has interpreted it;

> "A unified science in a unified language is really nonsense, I am sorry to say; and demonstrably so, since it has been proved, by Tarski that no consistent language of this kind can exist. Its logic is outside it."[84]

However, apart from dealing another death blow to the project of classical empiricism (i.e. to Positivism in its various forms) Tarski's truth theorem is of far greater import to epistemology in general since it formalizes a long suspected fact; that systems of epistemology are as incompletely axiomatizable as are the foundations of mathematics and first order logic. The theorem effectively explains the repeated failure of classical Western system-building philosophy. These systems failed because they were erected on weak (non-constructive) epistemological foundations. Consequently, such projects are undermined by the very *constructive* logic of indeterminacy they seek to repress. This fact was intuited by Pyrrho at the very (Aristotelian) origin of this tradition. As such his system – and to a lesser extent those that were either directly or indirectly influenced by him, notably the Kantian system – is the only one which is still fully viable two and a half millennia later.

This in a sense "negative" achievement of Tarski's did not however prevent him from proceeding to propose a solution to what amounted to a crisis in *classical* logic and epistemology. This solution is Tarski's famous *Semantical* conception of truth. To avoid the problem of inherent inconsistency in natural language Tarski proposed the construction of a formal *meta-language* in which to assess the truth-value of statements expressed in the natural or *object language*. *Truth* in this new language was effectively defined as *correspondence to the facts*. That is; a statement in the object language is defined as true in the meta-language *if and only if the statement corresponds to the stated facts.*
Although Tarski's idea is simple and even banal (this is not Tarski's fault, truth is nothing if not trivial), its expression in a precise and formalized language (effectively inspired by Gödel) *does* succeed in eradicating the problem of inconsistency in purely natural language that Tarski had demonstrated in the *undecidability of truth* theorem.
However, in my opinion, Tarski merely succeeds in shifting the burden of the epistemological problem *out* of the natural language and into the *formal* language. What effectively happens is that in solving the problem posed by the *second* incompleteness theorem to natural language Tarski merely succeeds in triggering the *first* incompleteness theorem as it applies to *formal* languages such as his own *meta-language*. The object language is *purified*, but only at the cost of fatally tainting the meta-language – a disaster, since the meta-language is now charged with carrying the burden of the object language as well! And if the meta-language falls foul of the first incompleteness theorem then this leads to a collapse in the claim to completeness of the object language. This claim can therefore only be maintained artificially.

---

[84] Karl Popper, *The Demarcation between Science and Metaphysics*. Reprinted in *Conjectures and Refutations,* (Op cit.) Section 5, p269.

Thus incompleteness and undecidability infect ideal languages as well, severely reducing their usefulness to philosophy. It is for *this* reason that Tarski's semantical theory of truth is referred to as a *theory* and not as a *theorem*.

Tarski's maneuver to avoid the *undecidability of truth* problem is, in essence, akin to that of those physicists who seek to evade the indeterministic (in effect *constructive*) implications of the uncertainty principle by assuming that – provided we do not directly interfere with it by measuring it – a particle *does* have a determinate position *and* momentum. But this is simply an error of classical deterministic thinking as has been demonstrated by experimental violations of Bell's inequality. To put it another way; if there *are* no *deterministic* facts (which quantum mechanics demonstrates to be the case) then how can a statement *correspond* to them (i.e. to what effectively does not exist) in a fully determinate and wholly unproblematic fashion? In essence Tarski's *Semantical* theory (implicitly adopting Aristotelian *essentialist* ontology[85]) assumes that facts *must* be determinate. The theory provably fails because this assumption is not sustainable.[86]

What this effectively proves is that the *first* incompleteness theorem applies to natural language just as surely as does the second, a fact which should in any case be intuitively obvious, but does not seem to have been, either to Tarski himself or to such prominent followers such as Karl Popper and Donald Davidson[87]. Thus, just as the two incompleteness theorems apply decisively to mathematics as a bearer of complete and consistent truth[88] so also they apply (by logical extension) to *natural language* as well which, as a bearer of truth, is not merely inconsistent but irretrievably incomplete (because *constructive*) as well. This (indirect) demonstration of incompleteness amounts to the conclusion of what might fairly be called the *second undecidability of truth theorem*.

---

[85] An ontology which is in many ways the basic assumption of classic *Realism*.

[86] The viability of Tarski's theory and the indeterministic implications of quantum mechanics seem to me to be inextricably linked issues. Nevertheless it is the (non-constructive) *logic* of the theory which is primarily at fault, as I have sought to demonstrate in this section.

[87] In effect Davidson's famous theory of meaning is fatally undermined by virtue of its dependence on Tarski's demonstrably incomplete truth theory. (Incidentally, if the truth theory *were* complete it would be called a truth *theorem!*) The alternative and correct view concerning the nature of meaning is that it is *relativistic* and so transitory in nature. This view is similar to, but more general in its expression than the Wittgenstein-Dummett view which is that "meaning is use". This view, I would say, is merely a logical corollary of the relativistic interpretation which implies that meaning, like time and space in Einsteinian physics, is a *perspectival* rather than an absolute phenomenon. Meaning, in other words, is relative to specific frames of reference; i.e. contexts. This is not the same as to call meaning "subjective" any more than time and space may be called subjective under relativistic mechanics. However, being a more complex phenomenon meaning is not reducible to simple mechanics either. It is, we might say, an emergent product of complexity, as, eo ipso, is ethics.

[88] This is presumably *because* mathematics is a *constructed* phenomenon, in the Intuitionist's sense, rather than a *Platonic* noumenon.

In point of fact so called *non-constructive* or *classical* logic is *also* Constructive in nature since Intuitionistic logic retains classical logic as a special case. It is only the (fallacious) law of excluded middle itself which is "non-constructive".

53. Neo-Logicism.

The Aristotelian law of excluded middle[89] and its main corollary – the principle of contradiction – are assumptions which hold within Tarski's theory of truth, but are abandoned as arbitrary in Heyting's at once simpler and more general (rationalized) system of Intuitionistic logic.

This suggests that the two theories are antagonistic and in rivalry with one another to form the epistemological basis of knowledge. In one sense this is correct, with Tarski supporting an essentially Aristotelian (or Chrysippian) view and Heyting an implicitly Pyrrhonic one. In practice however Heyting's logical system (the logic that incorporates indeterminism) retains the deterministic schema of Aristotle and hence of Tarski (which are bivalent) as a special case of itself. The correctness of this interpretation has been confirmed by Glivenko.

Additional confirmation is supplied by the fact that the continuum hypothesis of Georg Cantor has been demonstrated (by Kurt Gödel and Paul Cohen) to be inherently undecidable.[90] This constitutes proof that the *law of excluded middle (i.e. tertium non datur)* cannot be of universal validity and that therefore non-classical logic must constitute the correct epistemological basis (since it constitutes the only other possible alternative).[91] This, as mentioned earlier, amounts to a revival of Logicism, albeit on a wholly new (non-Aristotelian) basis to that conceived of by Frege.

Andrei Kolmogorov devised, around 1925, the means of translating classical proofs into intuitionistic or constructive ones, thereby indicating that Hilbert's understandable fears (echoing Aristotle himself) concerning the abandonment of *tertium non datur* are unfounded. Constructive mathematics is not an alternative or irrational form of mathematics, simply a more *general* and inclusive one. It is, incidentally, the correct basis of mathematics from the point of view of neo-empiricism. Intuitionism did not therefore portend the end of mathematics as Hilbert feared, but it did augur the end of Hilbert's classical formalist project, which received its final death blow at the hand of Gödel. Non-classical logic could however conceivably be employed as the correct basis for a *Neo*-Formalist project. In this scenario the ZFC system could formally be described as a *special case* of a non-classical logical schema. In this instance consistency and completeness could be restored, but at the cost of admitting *indeterminacy* as an objective category in its own right.

It is, incidentally, highly suggestive that the whole of determinate, classical mathematics can be deduced from the empty set, as it were, just as the whole of empirical science can be accounted for ex-nihilo. This is suggestive of the ultimate common nature and significance of both analytical *and* synthetical knowledge. This of course was a major tenet of the logical positivists.

An additional problem besetting classical formalism however is what might be called the *undecidability of the empty set*. Set theorists have sought to derive the whole of mathematics from the empty set (a kind

---

[89] Also maintained by Chrysippus as the epistemological basis of the Stoic's system.

[90] A similar point can be made concerning the inherent undecidability of the *axiom of choice* which can neither be proved nor disproved by the other axioms in the ZFC axiom system and so is independent of them.

[91] Although it is true that there *are* many different forms of constructive logic it is nevertheless correct to say that they all share one thing in common; the rejection of the law of excluded middle. Thus it is fair to speak of constructive logic as a single, albeit voluminously rich, entity.

Of *trivalent* logic for example Lukasiewicz has written; "If a third value is introduced into logic we change its very foundations. A trivalent system of logic, whose first outline I was able to give in 1920, differs from ordinary bivalent logic, the only one known so far, as much as non-Euclidean systems of geometry differ from Euclidean geometry. in spite of this, trivalent logic is as consistent and free from contradictions as is bivalent logic. Whatever form, when worked out in detail, this new logic assumes, the thesis of determinism will be no part of it"
(Jan Lukasiewicz, "*On determinism*" [1946], in *Selected Works*, North-Holland, Amsterdam 1970 (L. Borkowski, ed.). p. 126).

In point of fact, as Glivenko has demonstrated, Lukasiewicz was wrong in assuming that determinism has no part to play since it infact constitutes a special case of constructive logic.

of mathematical *creatio ex nihilo*) and yet its validity is objectively undecidable. Undecidability, in a way not hitherto fully grasped, constitutes a form of direct *verification* of constructivist assumptions concerning tertium non datur, as well as of neo-logicism. Analytical philosophy must therefore deal with the consequences of this situation, and it is much of the purpose of this work to show how this should be done. This undecidability, however is only a problem for the classical formalists and is easily coped with by non-classical logic.

The objection that could legitimately be raised against what might be called *Neo-Logicism* is that, in principle, constructive and many-valued logic do not necessarily include nor exclude any possibilities, thereby hypostatizing the problem of decidability. For the empirical sciences however Popper has already solved this problem as it pertains to synthetical statements.[92] Indeed this problem *can* only be solved on a constructive basis, a connection Popper (tied as he was to a commitment to Realism and to Tarski's incomplete theory of truth) sadly never made. Constructivism is clearly the only viable epistemological basis for synthetical knowledge and attempts to rely on classical logic (notably in the form of logical-positivism) have demonstrably failed.

For analytical knowledge the case is equally clear cut. Gödel's incompleteness theorem and Church's theorem concerning the undecidability of first order (i.e. classical) logic effectively demonstrate a *constructive* epistemological basis to both logic and mathematics, as do the points made previously concerning the continuum hypothesis and the axiom of choice.

These facts indicate the need to understand the Zermelo-Fraenkel axiom system (which is the epistemological basis of all *decidable* mathematics) within a *constructive* context supplied by Intuitionistic logic. In doing this we effectively render the foundations of mathematics (notwithstanding undecidability) logically complete. In essence, anything which cannot be accounted for within classical mathematics requires reference to non-classical logic in order to maintain rational, albeit (what I term) non-classical completeness. Furthermore (just to amplify the point about completeness) non-classical logic retains classical logic as a special case. Thus all the happy felicities of classical logic and mathematics are retained and the incomparable (and currently absent) benefit of rational completeness (with respect to the logical basis) is obtained as well.

All *decidable* mathematics is dependent on the Zermelo-Fraenkel system of axioms which – due to Gödel's theorem – is not completely decidable in its own right and is therefore implicitly dependent on constructive logic. *But any questioning of the decidability of constructive logic **itself** presupposes constructive logic and therefore we say that constructive logic is complete.* And herein lies the epistemological benefit of adopting non-classical logic as our rational foundation.

Consequently, rather than the usual *negative* interpretation that is given to the theorems of Gödel and Church - that they demonstrate the incompleteness of *classical* mathematics and hence the failure of formalism and logicism alike - what I propose instead is a *positive* interpretation; that they implicitly demonstrate the correctness of *Constructivism* (in mathematics and logic alike) and the completeness of logic and hence mathematics with respect to *Constructivist* (i.e. non-classical, but still fully rational and formal) assumptions. The incompleteness theorems and their derivatives are thus the gateways to constructivism and so are not characterisable as negative results at all (except from a narrow classical perspective).

The essence of the neo-rationalist philosophy that arises as a consequence (and I would say a *special case)* of neo-logicism is that an *absolute* distinction between determinate and indeterminate knowledge (be it analytical or synthetical in nature) cannot be maintained. Instead, what we justifiably distinguish (either empirically or analytically) as *determinate knowledge* is, strictly speaking, a *special case of indeterminate knowledge*. This recognition, above all, allows a precise and **practical** distinction to be drawn between determinate and indeterminate knowledge, but *not an **absolute** one* of the type sought after by *classical* rationalists and empiricists alike. According to neo-empiricism *indeterminism* constitutes the general case, *determinism* the special or limiting one.

The problem concerning the inherent indeterminacy of *the given* in physics and of inherent incompleteness in mathematics and logic (which are at root the same problem) is effectively dissipated

---

[92] See Popper *The Logic of Scientific Discovery,* op cit.

by this interpretation. Furthermore, the bulk of the classical disputes that have arisen in philosophy and which constitute the distinctive "problems of philosophy" arise primarily from *this* epistemological state of affairs (i.e. from indeterminacy) and *not* from poor syntax and language use. In other words, the problems of philosophy primarily arise because although syntax *is* logical the logic it follows is *non-classical* or polyvalent, thus allowing or making space for the ineluctable presence of indeterminacy. Indeed non-classical logic may be characterized *as* the logic of indeterminacy. Without non-classical logic neither language nor philosophy would be able to cope with the fundamental ontological and epistemological fact of indeterminacy.

The point also needs stressing that non-classical logic, as I conceive it, does not require the construction of an *"ideal language"* a la the logical positivists etc., but can be applied, in an informal manner (or just as an implicit theoretical perspective) in the context of ordinary language. It is, in other words, the natural logic presupposed by even the most *informal* species of linguistic analysis.

Furthermore, we might fairly call the *bivalent* "ideal" language of the classical positivists a *special or incomplete case* of the multivalent *real* or "ordinary" language. But there is little real purpose served by this highly formalized "special case". However, *were* the logical positivists to admit of additional truth values to their systems then they would no longer be positivists. Hence (for example) their avoidance of Popper's interpretation of the logic of induction, even though this analysis is both logically correct *and* comparatively simple.

Of course what is called *ordinary language philosophy* is *also* changed by this interpretation since the notion that philosophical problems, with their roots in indeterminacy, can always be solved or eradicated simply through linguistic analysis *also* presumes a fundamentally bivalent and deterministic view-point, one which is not, I think, justifiable. This is because it proves impossible, in practice, to define a clear and objective *demarcation* of the boundary between real and pseudo-problems. Indeed, such a boundary is *itself* indeterminate, a fact not grasped by the ordinary language philosophers who currently hold sway in academia. As a result the approach of linguistic analysis, though valuable in itself, cannot be viewed as the universal panacea it purports to be.

As we have already seen, even Popper's principle cannot be absolutely definitive in providing demarcation (between "real" and "pseudo" problems), even when used with hind-sight. It is for example *only with hindsight* that we are able to see that the approaches taken by Parmenides and Democritus were not pseudo-solutions to pseudo-problems. Thus the attempt to handle the problems created by indeterminacy purely by means of linguistic analysis can be seen to be inadequate.

What Austin has called the *fallacy of ambiguity* is thus not the true fallacy. The true fallacy is the belief that ambiguity *can* be eradicated, either by the methods of the ideal language philosophers *or* by the ordinary language school of which Austin was a part. This we might reasonably call *Austin's fallacy*. The solution to this erroneous perspective is simply the correct analysis and understanding of the concept and application of indeterminacy offered (for example) by this work.

Non-classical logic, it therefore follows, is *prior even to language* and represents the *apriori* basis upon which the analysis of language (and thus thought) proceeds, whether we choose to recognize it or not. This, indeed is our fundamental hypothesis concerning the nature of philosophy as a distinct and enduring discipline.

As such, it needs to be stressed that non-classical logic *must* be interpreted as constituting the true and distinctive foundation of *all* philosophy (of mathematics and physics alike), one which, when clearly grasped, leads to the all important recognition of indeterminacy as a *logical* and not merely an *empirical* category (although it ultimately supplies the point of profound unity between the analytic and the synthetic).

And it is primarily in this *broader* sense that I use the term *neo-logicism*, although the narrower sense (referring only to the foundations of mathematics) also applies as a particular case of neo-logicism. My point concerning empiricism is that since its method is fully logical in nature (given Popper's analysis) then this logic must be seen as a part of a broader treatment of logical foundations which incorporates not merely analytical knowledge but (as a special case) empirical knowledge as well.

Since the logic of science is analytical then it is also, ipso-facto a part of a *broader* logical scheme of knowledge. And so the clear division between the analytical (which is broad) and the synthetical (which is narrow) is seen to fail. This is because, as I will demonstrate, the synthetical, *in every respect*, is *analytical* in disguise. What this surprising hypothesis means is that a complete treatment of the analytical is ipso facto a complete treatment of the synthetical as well. Ergo, neo-logicism incorporates an account of *both* species of knowledge. And this, incidentally, explains the strange and controversial results of Kant (that the synthetic is apriori) and of Quine (that the synthetic and the analytic are really one).

54. What is Truth?

It is important to remind ourselves that Aristotle's view of the matter (that; "Truth is to say of what is that it is and of what is not, that it is not,"[93]) is not wrong per-se, but only incomplete. The same is true, for example, of Tarski's theory of truth or Popper's realism. And the fundamental reason for this state of affairs, as I have sought to argue, is that indeterminacy is not merely an epistemological category but an ontological one as well.[94] If (after quantum theory) the very *ontological* reality of *the given* is open to question then so too must be its epistemological status – which is indeed what we find. It is this demolition of the idea of *givenness* (what Derrida later calls the "metaphysics of presence") which renders the realist position not untenable but undecidable.[95] Constructive logic – which is a fully recursive form of logic – simply reflects this state of affairs.

Truth is inherently and objectively indeterminate therefore, due to the condition I have called *Universal Relativity*. As a Tarskian *meta-statement* we might even say that *this* is what truth *is*. But this (though "true") is an empty statement. What it means is that, at all times, we hold to a relative (or contingent) sense when we use the word *truth*. It ought to be added that the same set of points can be made concerning the correct use of the word *false*. The "meaning" of these two words is, in every sense *constructive* and relativistic.

Given the category of *undecidability* it follows that the nature of truth cannot be definitively ascertained, one way or the other, a fact which undermines the epistemological assumptions of instrumentalism (which argues that scientific theory has *no* inherent truth value and is merely a tool for making predictions), just as it does those of realism. Both are equally guilty of dogmatism since the question of truth-value is inherently undecidable.

The concealed dogma of various forms of Pragmatism (for example instrumentalism or else so called *naturalized epistemology*) lies in the unsupported *denial* of a truth value to statements, whose significance is instead judged solely in terms of their practical consequences. The basis for this dogmatic judgment lies merely in the failure of the realists to prove *their* case. The failure of realism, according to the Pragmatists, *proves* the validity of instrumentalism. But infact it does not, it merely supplies corroborating evidence in *support* of their thesis. The failure of realism proves nothing except the failure of realists to prove their case.

Realism, in the light of arguments propounded by Popper, Tarski and others (e.g. Davidson) is nevertheless at least as tenable as instrumentalism – which is precisely the point. The problem of Truth is simply not decidable as between the two camps, whose arguments and counter-arguments lead to an infinite regress which is *inherently* undecidable. This is why I take the view that the logically correct position on the issue is to accept its inherent *undecidability*. This is the case, indeed, with most if not all specifically *philosophical* questions. These questions are *not* usually meaningless (as according to the Carnap-Wittgenstein interpretation of them) nor are they in want of further elucidation, as if additional information might resolve the issue. Nor yet are they simply pseudiferous (since the line of demarcation between pseudo and legitimate problems is not determinate).They are simply and inherently *undecidable*. They are also infinite in number.

The validity of Instrumentalism cannot, by definition, be proven and therefore its dogmatic assertion (such as it is) is unsustainable from a purely logical stand-point. Pragmatism, though it may contain an

---

[93] M G 7 1011 b 27

[94] As the reader will recall from our work on the physical picture of the universe *indeterminacy* has a precise mathematical definition; zero divided by zero.

[95] See also Michael Dummett (1963). *Realism,* reprinted in: Truth and Other Enigmas, Harvard University Press: 1978, pp. 145-165.

Dummett's error, like that of other instrumentalists, is that of a sort of *dogmatic* anti-realism. Since the issue is undecidable and somewhat linguistic in nature it is technically incorrect to be dogmatic on either side of the issue. The best analogy is perhaps with Paul Cohen's mathematical proof that the continuum hypothesis is both true *and* false.

element of truth, is nothing more than an act of faith – indeed, Realists could argue that Pragmatism is self refuting since it implicitly asserts that; the denial of truth-values is in some (second order) sense "true".

Conversely, those forms of Pragmatism which identify truth with utility do, like Utilitarianism before them, expose themselves to charges of pure subjectivism, since they merely beg the question as to what constitutes or determines "usefulness". On this view *everything* may be true one moment (because it is "useful" in that moment) and nothing may be true the very next moment (since it has suddenly stopped being useful), which, I would have thought, is the very acme of absurdity. Furthermore, what could be more "useful" than the (polyvalent) logic of induction? Yet it is precisely this form of logic which indicates that the issue of *truth* is not decidable. All of which strongly points to the inherent *undecidability* of the issue of truth, which, needless to say, is the correct *logical* stance (allowing us to transcend the twin dogmas of realism and pragmatism) on this particular issue.

55. Naturalized Epistemology and the Problem of Induction.

Pragmatism in many respects represents yet another response to Hume's problem of induction, a problem which has remained at the center of epistemology for the last quarter of a millennium. But neither it nor Kantianism nor Positivism has succeeded in solving the problem since its solution lies in a form of logic not recognized by them.

The problem of induction should rather be interpreted as indicating *incompleteness* in the underlying logic of empiricism and so clearly represents a problem of epistemology as classically conceived. The problem of induction, *as* a logical problem is (as even Popper failed to realize) simply a sub-instance of the wider problem of undecidability in first order logic (as demonstrated in Church's theorem) which is in turn a sub-instance of the problem of incompleteness (and inconsistency) that is highlighted in Gödel's two incompleteness theorems. As such the problem of induction (since it is ultimately a problem of logic) should be interpreted as a special case of the problem of *decidability* in mathematics and indeterminacy in physics. This, we may say, is therefore the fundamental problem of epistemology and ontology alike.

Although Quine quite rightly recognizes the impossibility of finding a solution to the problem of induction on a Positivistic basis (and this, we now know, is due to the latter's reliance on bivalent logic), Quine's alternative solutions (first *Confirmational Holism* and later *Naturalized Epistemology*) are, if anything even less successful. For although Positivism is an incorrect response to the problem of induction (which arises as a *consequence* of Positivism) it does not, at any rate, represent a category mistake.

In truth the only valid and rational treatment of the problem of indeterminacy is through non-classical logic which should therefore be seen as representing the foundation of epistemology. Acceptance of this hypothesis (whose justification, since it lies in the *recursive* quality of non-classical logic, is therefore *apriori* in nature) obviates the need for big Metaphysics at a stroke.

What appear to be problems of Metaphysics are therefore found to be problems of (non-classical) logic in disguise. It is only in this sense that Kant's intuition (concerning the apriori foundations of epistemology) may be found to be valid. Thus all the problems of metaphysics can be outflanked with reference to constructive logic and (by natural extension) mathematics and the sciences.

The value of logic over all other types of metaphysics lies in the matter of its precision. Thus when faced with a choice concerning which type of metaphysics to use at the foundations of our system we should employ Ockham's razor so as to prefer non-classical logic ahead of Phenomenology or any other type of metaphysical solution. Nevertheless we should remember too that logic is *also* metaphysical in nature and indeed (in my view) stands at the foundation not only of the sciences but of metaphysics as well. Thus whenever metaphysical arguments are found to be valid (in Phenomenology for example) they may *also* be found to be translatable into the language of (non-classical) logic. Peering through the miasma of metaphysics (as it were) what one discerns is the crystalline clarity of apriori (i.e. non-classical) logic.

In contrast to this, so called Replacement Naturalism, like Pragmatism before it, represents little more than an out-right rejection of the very possibility of foundationalist epistemology as an independent subject. In the light of the successes of Constructive logic it is now apparent that this rejection of the very possibility of traditional epistemology (a rejection which is implied by Pragmatism per se) is a good deal premature.

Furthermore, Quine's attempted conflation of science and epistemology seems itself to constitute a *category mistake* on a scale unmatched since Kant's attempt to solve the problem of induction (a problem of logic) through recourse to a vastly overdetermined metaphysical system. Quine's strategy (given the implied failure of confirmational holism) involves a de facto abnegation of epistemology and hence (thereby) of Hume's central problem. This strategy, which advocates the displacement of epistemology by psychology, is driven by a pessimism that is itself fuelled by the unsustainable

Aristotelian dogma of bivalency which states that epistemological foundations *must* be *either* certain *or* non-existent. Since, as Quine correctly observes in his essay, Hume's problem has proven insoluble in the hands of Positivists such as Carnap, he erroneously draws the conclusion that the traditional search for epistemological foundations (prior to and apart from empirical science) must therefore be hopeless;

> "The old epistemology aspired to contain, in a sense, natural science, it would construct it somehow from sense data. Epistemology in its new setting, conversely, is contained in natural science, as a chapter of psychology… The stimulation of his sensory receptors is all the evidence anybody has to go on, ultimately, in arriving at his picture of the world. Why not just see how this construction really proceeds? Why not settle for psychology... Epistemology, or something like it, simply falls into place as a chapter of psychology and hence of natural science."[96]

Logic is not however an empirical science (since it is apriori in nature). Indeed, it exhibits *far greater generality* than do the sciences and (given Popper's analysis) even seems to underpin them. Therefore Quine's conclusions cannot be correct.

Quine's solution therefore amounts to saying that logic is "contained" by natural science, which is certainly absurd. Indeed it highlights a tendency in Quine's avowedly *empiricist* philosophy to downplay the importance of logic as an independent organon in favour of what he dismissively calls "surface irritations" (i.e. sensory experience) as the sole source of knowledge. The "logic" of this is indeed an epistemology such as *replacement naturalism*.

Popper has however supplied the only valid form of logical empiricism, one which, unlike Positivism is not dependent upon the discredited principle of bivalency. By proving that logical empiricism is indeed still viable Popper has effectively contradicted the views of Quine as expressed in the essay I have just quoted from and thereby rendered Quine's alternative to logical empiricism redundant.

Furthermore, Quine's alternative continues to beg the question as to the *rational* basis for empirical psychology and for its alleged epistemological insights, thereby, in effect returning us back to where we began – i.e. to the (admittedly *non-classical*) problem of induction. Though precisely what the epistemological insights of naturalized epistemology *are* is still unclear, some thirty years after Quine started the Naturalistic ball rolling and also some hundred years after Dilthey, Brentano and Husserl started the surprisingly similar ball of Phenomenology rolling. We therefore rather suspect replacement naturalism (not to mention Phenomenology – which promised so much and delivered so little) to be merely a means of conveniently shelving what is, after all, an extremely irritating problem (the problem of induction and indeterminism), albeit not a surface irritation.

---

[96] Quine, *Epistemology Naturalized*, 68, 75 and 82, printed in *Ontological Relativity and Other Essays,* New York, Columbia University Press, 1969.

56. The Problem of Epistemology.

It is clear that Quine and the empiricists are unduly impressed by the *"isness"* of things. But it is also clear that the universe is a *logical* phenomenon first and an empirical (and hence psychological) phenomenon only incidentally to this. For after all, if logic dictated that there "be" *absolute nothingness*, then there would *be* no empirical phenomena (or subjectivity) in the first place. Fortunately *apriori* logic dictates the *exact opposite* of this. Thus we may conclude that empiricity is *contained* in rationality (or analyticity), and not (as empiricists seem to think) the other way around.

It is therefore ironic that the problem of induction – in essence a problem of logic – has continued to remain open many decades after its undoubted solution by Popper. As a result, progress in epistemology has all but ground to a halt, in favour of the erroneous phenomenology of Quine and Husserl. However, the mere failure of *bivalent* logic to solve this problem does not justify recourse to the "solutions" offered by the Phenomenologists.[97] Phenomenology (including Quine's distinctively empiricist brand of Phenomenology) may therefore be said to be redundant since it exists to counter a problem that has already been solved. A similar judgment should also be passed on Kantianism.

The convenient relegation of the work of Popper to the so called "philosophy of science" doubtless has much to do with this sorry state of affairs in contemporary philosophy. And yet epistemology since at least the time of John Locke has, in large part *also* been dismissible (in principle) as mere "philosophy of science". For the truth remains that epistemology (given the problem of induction) has been driven largely by issues arising from the sciences. Hence, no doubt, Quine's infamous quip that "philosophy of science is philosophy enough".

The most famous example of the direct influence of the problem of induction on subsequent philosophy is Kant's system, a system which may be said to be the progenitor of phenomenology. Thus the problem of induction might fairly be viewed as more seminal than any of the many assays that have been made upon it, not even excluding Kant's enormously influential system.

Of course Kant's system does more than simply address the problem of induction, though this problem is at the heart of what drives it. Kant seeks to explain why, notwithstanding the problem of induction, we are yet able to share inter-subjective (what he calls "objective") experience of the world, a fact which Kant, correctly in my view, attributes to the existence of an *apriori* element to experience. Properly developed this underpinning idea may be said to lead not to Phenomenology, but, more properly, to the revolution in logic inaugurated by Frege and developed subsequently by the intuitionists and the Warsaw school of logicians.

Kant appositely explains his own mission thus wise;

> "But I am very far from holding these concepts to be derived merely from experience, and the necessity represented in them, to be imaginary and a mere illusion produced in us by long habit. On the contrary, I have amply shown, that they and the theorems derived from them are firmly established a priori, or before all experience, and have their undoubted objective value, though only with regard to experience…
> Idealism proper always has a mystical tendency, and can have no other, but mine is solely designed for the purpose of comprehending the *possibility of our cognition a priori as to objects of experience*, which is a problem never hitherto solved or even suggested."[98]

---

[97] I prefer to call them category mistakes rather than solutions.
[98] Kant, *Prolegomena,* section six. Emphasis added.
 As an idealist Kant believed that all cognition is, by definition, subjective in nature. His breakthrough lies in seeing that subjectivity is *itself* constructed out of certain apriori elements (what he calls the categories and forms of intuition) and is thus identifiably objective or "transcendental" in character. In essence therefore he demonstrated that the absolute cleavage between subjectivity and objectivity is unsustainable.

In essence I am broadly in agreement with the conclusions of Kant's system on these matters, although advances in physics and in logic allow us to come to a similar conclusion with more precision than was available to Kant. The problem of Idealism (i.e. of the subject) underlying Kantianism and Phenomenology, for example, ceases to be significant because the apriori element turns out to be *analytical* rather than phenomenological in nature.

In other words, the aprioricity of logic remains *unaffected* one way or the other by the issue of subjectivity. The issue of subjectivity may therefore be said to be *subordinate* to that of logic and so may be relegated (along with the whole approach of phenomenology) to mere (empirical) psychology.

But whilst Kant primarily had the problem of induction in mind I also have the related, but more *general* problem of *indeterminacy* in mind, of which the problem of induction is, in many ways, but a sub-instance. *Principia Logica* seeks to explain not merely how order comes about *inspite of* indeterminacy, but *also* seeks to show, (relying not on linguistic verbalism, but *solely* on logic and empirical analysis,) how order and objectivity are made possible precisely *because* of indeterminacy. I flatter myself that this too is a problem "never hitherto solved or even suggested."

It would therefore be an exaggeration to say that "Epistemology of Karl Popper is epistemology enough", but only just. This is because the problem of induction as a problem of *Logic* is ipso-facto a sub-instance of a *still broader* problem (which, even more than induction, should perhaps be considered as *the* problem of epistemology) which is that of incompleteness and undecidability in classical mathematics and logic.[99] The solution to *this* mother of all problems in the theory of knowledge lies, as I have already discussed, not in Phenomenology etc, but in the adoption of *Constructive* logic and mathematics, which are recursively consistent.

---

[99] Clearly, in this case, philosophy of science is *not* "philosophy enough", but lies contained (in the form of Popper's solution) as a special case of the *philosophy of mathematics* which, who knows, possibly *is* philosophy enough.

57. The Correct Interpretation of Instrumentalism.

Although Instrumentalism and Pragmatism do not work as epistemologies in their own right they do nevertheless supply valid *responses* to epistemology. That is; since truth is undecidable it therefore makes sense to utilize hypotheses (of every sort) in an instrumental or pragmatic fashion. In other words, instrumentalism may not be a viable epistemology in its own right (any more than realism is) but it is a viable *consequence* of a logically valid epistemology. This use of instrumentalism makes no presuppositions about truth values (and so is not logically suspect), but, at the same time, neither does it qualify as an epistemology either. This, indeed, is the correct logical form and status of instrumentalism.

Although scientific theories *can* be used purely as instruments for making predictions it does not follow from this that scientific theories are also excluded as potential bearers of truth. Such an exclusion amounts to mere dogmatism since only falsification of a theory can rule out this possibility. Furthermore, thanks to the validity of the Quine-Duhem paradox, even falsification of theories does not constitute a definitive judgment on them. Hence the need to place *indeterminacy* at the heart of our epistemological (as well as our ontological) understanding. Decidability, where it *does* occur, can (because of the Quine-Duhem paradox and because of Godelian incompleteness) only ever be *contingent* in nature.

Nevertheless, *a contingent decidability vis-à-vis empirical and analytical statements* is, as discussed earlier, a vitally important phenomenon for *practical* purposes. Indeed it is this form of knowledge which we generally refer to as "true". But the fact remains that such knowledge is merely a contingent (contingent, that is, on *uses, contexts and definitions*) and special case of *indeterminacy*, which latter therefore represents the *more general state of affairs*. From this it follows that any successful determination of any given problem of knowledge or philosophy is always *purely relative* in nature. There can never be such a thing as an *absolute determination,* only a determination relative to a given set of pre-suppositions. Hence, we suppose, the *persistency* (and indeed the *infinity*) of the perceived problems of philosophy.

These problems arise because we tend to think of determinacy as the norm and indeterminacy as some kind of irritating limit phenomenon which it is the purpose of philosophy and the sciences to eliminate once and for all somehow. Nothing could be further from the truth however, hence the failure of most classical philosophy. In truth, *indeterminacy* is the general case, determinism the limit phenomenon. And thus we say that the true logical foundations of philosophy (which it is the primary function of philosophy to gauge) are non-classical. Indeed, philosophy is nothing more than the recognition of this general state of affairs.

Indeterminism is also responsible for the sense, first detected by Quine, of the *under-determination* of our theories. In truth however, it is more logically correct to say that our theories are neither *under* nor *over* determined, they are simply *indeterminate* through and through, even though, *on a purely relative basis,* (as we have just discussed), we may sometimes say that our theories *are* determinate (contingently speaking) as well.

It also follows from our discussion of the rival claims of instrumentalism and realism that Kant's famous distinction between phenomenon and *ding-ans-sich* ("thing itself") far from being the unproblematic tenet of modern philosophy that it is usually assumed to be, is in reality, logically *undecidable*. With it goes much of the logical basis of Kant's system and that of his successors.

Conversely, the *identity* of phenomenon and thing itself (such as is to be found in certain naïve forms of classical empiricism and positivism) is not decidable either.

58. The Roots of Phenomenology.

It was Schopenhauer who pointed out a potential loophole in the simple dichotomy between phenomenon and *thing itself*. Since the Cartesian subject exists and yet is not a *phenomenon* as such (at least not unto itself) it therefore seems to follow that, on Kant's reckoning, the subject must be an instantiation of the *thing itself,* or *noumena*. And since it *is* possible for us to "know ourselves", as it were, then ipso facto, it is possible, contra Kant, to directly know the ostensibly unknowable *noumena.*
Although this reasoning represents the origin point of what subsequently became the *dominant* strain of post Kantian Idealism I prefer to interpret it as evidence of the correctness of my view concerning the inherent *undecidability* of this classical distinction. Indeed, Kant's famous distinction is one of those rendered irrelevant by a correct logical understanding of the concept of indeterminacy. And with this distinction would also go, ipso-facto, much of post-Kantian philosophy. Since much of what was to become Phenomenology flowed from Schopenhauer finesse of Kant's dogmatic distinction (and similar maneuvers by Hegel) it seems to me that, logically speaking, Phenomenology is severely *compromised* rather than aided by this fact.

According to Husserl Phenomenology reduces to the study of the stream of consciousness in all its richness, out of which may be distilled an awareness of apriori determinants, physical laws and inter-subjectivity. But all of these are knowable only subjectively *as* phenomena. Thus Phenomenology distinguishes itself *in no way* from empiricism other than in its flight from precision and into introspection. As it were, an equal and opposite bias to that of the Positivists.
We may however remind ourselves of my criticism of the absurdity of treating logic as if it were a *phenomenon,* when in truth it is apriori. Nothing gives away the empiricist roots of Phenomenology more than does this error.
Phenomenology is also partly reducible to transcendental idealism, since Kant intended his system to be rooted in an awareness of the inescapable centrality of the subject as the locus of all supposedly objective knowledge. We know the "subjectivity" of others, the "objectivity" of the sciences (including physics and psychology) only through the aforementioned stream of consciousness which is therefore the locus of our individuality as well as of our sense of what is held in common. All knowledge, in order to be truly authentic must, we are told, be grasped as existing in this context.
Thus Phenomenology leads effortlessly to existentialism with its focus on the individual as the twin center of all knowledge *and* subjectivity alike. It also leads ultimately to the intensely subjective style of philosophizing of Heidegger and Derrida. Nevertheless this entropic trajectory of Idealism has perhaps been latent in modern philosophy since the time of Descartes, with major milestones represented by Berkeley, Hume and Kant. Heidegger and Derrida are merely the ne-plus-ultra (one presumes) of this subjectivist tradition of modern philosophy, a tradition, incidentally, which cannot easily be refuted due to the validity of its initial Cartesian premise (which is that the subject is the locus of all our supposedly objective knowledge). Subjectivity, for Husserl, thus represents what he calls the *transcendental problem* whose solution lies in the recognition of what he calls the *relativity of consciousness,* in effect *transcendental relativism.* Thus universal relativity, according to the phenomenologist, is latent in Kant's system.
The assumption that logic is somehow compromised by the existence of subjectivity and that subjectivity (Husserl's mystical *ideen)* is therefore *prior* to all categories and concepts seems to me to be the main departure from Kant's view, which is that apriori factors are (by definition) prior to and ultimately responsible for the transcendental subject. Yet Phenomenology (after Husserl) eventually renounces Kant's view and renounces too any commitment to accounting for inter-subjectivity, an approach which makes philosophy seem pointless, trite and even hypocritical.

Probably the most significant points of Heidegger's originality are what might be called the *ontological turn* together with an associated deduction (somewhat downplayed by Heidegger) of onto-nihilism which, like Husserl's deduction of "transcendental relativism" does indeed seem to be a latent presence in Kantianism.

Indeed onto-nihilism follows like a shadow once the argument for "transcendental relativism" is granted. It was Heidegger's originality, to some extent echoed by Derrida (for whom the critique of the "metaphysics of presence" is a still more central concern), to intuit and to posit – albeit obscurely – this fact. Indeed, the underlying unity of phenomenology and empiricism is also further evidenced by Husserl and Heidegger's deductions of relativity and nihilism respectively. Nevertheless, for reasons which I shall supply in the next section I do not think that Heidegger's treatment of Leibniz's "ontological question" is entirely adequate, in terms of either style or substance.

We might also observe that it is through Heidegger and Derrida that the "linguistic turn" is finally taken by the Phenomenological tradition, thereby bringing it that much closer to the analytical tradition. The fact that Phenomenology comes late to the linguistic turn indicates that perhaps it is the more conservative of the two major traditions in post Kantian philosophy.

59. Ontological Commitment versus Hyper-Nihilism.

In *Two Dogmas* Quine proclaims what he calls the "myth of physical objects" and yet in later essays he may be discovered urging his readers on to an existential commitment to the substantial existence or independent "reality" of various forms of universals, (notably those of set theory and propositional logic). It seems, in other words, that physical objects do not exist, but abstract ones do. This I take to be the fundamental contradiction at the core of what might be called Quine's *metaphysical system* and of his rather Calvinistic doctrine of *ontological commitment*.

The fact that we use language as a practical tool does not commit us to a belief in the ontological reality of the things (be they universals *or* particulars) of which we speak. Indeed, ontological commitment to universals only becomes the logical necessity Quine says it is if we have *already* made a prior ontological commitment to the belief in particulars. But this latter is merely an act of faith, certainly one which Quine has no right to assume we have taken.

After all, if nature is without substantial or at least *determinate* being (as physics *and* logic alike seem to indicate) then the problem of universals and hence of ontological commitment simply does not arise. Since there are no determinate particulars so, it seems to follow, there can be no determinate universals either. Ergo the problem of ontological commitment cannot arise. The best that can be said for the issue therefore is that it is another one of those *intrinsically indeterminate* problems of philosophy, of which (we have established) there seem to be an infinite number.

Of universals the most interesting and pertinent are perhaps the natural numbers, although Popper has made the point that in a sense *all* words are universals, a theory he calls *the transcendence inherent in description*. I think that Popper's paradoxical observation is indeed correct and it leads in turn to his refreshingly clear and natural interpretation of the language of mathematics, a language, (that is), which must logically take part in this *inherent transcendence* as well;

> "In other words, mathematical symbols are introduced into a language in order to describe certain more complicated *relationships* which could not be described otherwise; a language which contains the arithmetic of the natural numbers is simply richer than a language which lacks the appropriate symbols."[100]

Consequently, mathematical symbols do not describe *things,* as such (in the Pythagorean sense), but rather *sets of relations*. It is by misunderstanding this distinction that the problem surrounding the interpretation of universals (and their reification) arises.

Wittgenstein seems to make a similar point about concepts and universals *in general* – i.e. that their meaning does not derive from anything *intrinsic* to themselves, but rather from their place in the overall (what Wittgenstein calls) "language game" – i.e. *relative to one another;*

> "When philosophers use a word – 'knowledge', 'being', 'object', 'I', proposition', 'name' – and try to grasp the *essence* of the thing, one must always ask oneself: is the word ever actually used in this way in the language-game which is its original home?
> What *we* do is to bring words back from their metaphysical to their everyday use."[101]

Quite why philosophers make a rigid (yet crucially *undefined*) distinction between everyday language use and metaphysical use does I must confess escape me somewhat. After all ordinary language use is *imbued* with logical constructions and geometrical concepts and yet logic and geometry are certainly, in some sense, metaphysical and apriori. The down-playing (albeit not the outright denial) of this

---

[100] Popper, *What is Dialectic,* Mind, N.S., 49. 1940. (Emphasis added).
[101] Wittgenstein, *Phil inv,* op cit. Part one, section 116.

uncomfortable fact (which is after all also Kant's basic point) is indicative of the ultimately *empiricist* bias of ordinary language philosophy as of Phenomenology and contemporary analytical philosophy in general.

It seems to me therefore that the ambition to eradicate metaphysics is a misguided ambition since the demarcation between the physical and the metaphysical is inherently indeterminate and would in any case lead to the eradication of logic and hence of rationality itself. One can eradicate the latter concepts only at the expense of eradicating the concept of being itself.

One of the reifications that *is* undermined by this *relativistic* interpretation of meanings and of concept formation is the doctrine of *Holism* (of which that of Quine-Wittgenstein is but a species). This is because holism *idealizes* the concept of unity or oneness at the expense of numbers other than the number one, even though it is apparent, on the above view, that numbers do not have *intrinsic* meaning, but only a *relative* meaning, relative, that is, to all the *other* numbers in the continuum. Thus unity, plurality *and even nothingness itself* are without intrinsic meaning or being and can only exist as concepts *relative* to one another, a point of view I term *hyper-nihilism*.

We may for example ask what it *is* that is "One", i.e. one what? Other than the concept itself it is hard to envisage anything to which such a purified "oneness" may refer and yet the concept of holism, devoid of substance, (pure concept as it were), is a quite meaningless abstraction and so should, I think, be opposed.

Another reification that disappears under analysis is, ironically that of Nihilism itself. However, this logical self-negation still leaves us with nothingness, and so represents, I think, a more consistent and defensible (not to say elegant) position than that of Holism.

In essence reification (of the Nihilistic or the Holistic sort) is avoided with reference to the doctrine of *Universal Relativity*. By adopting the perspective of universal relativity ontological nihilism follows (like a shadow) but without the inelegant reification of *absolute* Nothingness (Nothingness with, as it were, a capital letter) such as we find in Heidegger's impenetrable system.[102]

Of course this doctrine (of universal relativity) and the *associated deduction of ontological nihilism* which follows from it, is not original to me (though I am attempting to place it in the context of modern philosophy) but represents the fundamentally original and outstanding contribution of Buddhism to epistemology and ontology. Indeed, the idea can be traced back to the Buddha himself. Furthermore, the idea complements the underlying logic of Hindu Annihilationism, which represents, as it were, the pre-history of Buddhist metaphysics.

---

[102] Consider for example Heidegger's famously obscure aphorism; "das Nichten selbst nichtet" ("the Nothing self-annihilates") from, Martin Heidegger, *Was ist das Metaphysics,* 1926.

60. Non-classical Completeness.

Further arguments against the Platonist and in favour of what I choose to call the *relativist* interpretation of universals, notably the natural numbers, is supplied by constructive or non-classical mathematics. As our earlier discussions may have indicated Gödel's undermining of classical formalism has had the side-effect of also undermining the traditional authority of the *classical* concept of *proof* in mathematics. All but the most elementary proofs in mathematics rely on a chain of proofs which ultimately terminate in a basic set of principles known as *axioms,* such as those of Euclid. The most encompassing system of axioms that have been devised are those of the Zermelo-Fraenkel system which may be said to axiomatize all decidable mathematics, but are incomplete with respect to constructive logic. As my work has already indicated, those hypotheses which are *not* contained by the ZFC system are in turn incorporated into a larger logical schema based on non-classical logic (a simple rectifying procedure I call *neo-logicism).*

In other words, there are no sustainable grounds for separating off undecidable mathematics from the rest of mathematics other than on the basis of its (no longer problematic) undecidability[103]. Now if this undecidability becomes an allowable category in the underlying scheme of logic then it follows that mathematics becomes *complete* with respect to that scheme. Thus Gödel's result is rationally accommodated and a form of (non-classical) completeness restored. This indeed is the whole purpose in adopting a non-classical scheme of epistemology.

Furthermore, the inherent undecidability of ZFC itself ceases to be problematic as well on such a logical basis. It is only on the strict criteria of formalism (which is inherently classical) that mathematics might be said to be incomplete. Alter those criteria, as in neo-logicism, and mathematics instantly becomes complete with respect to the new logical base. Given the completeness of mathematics on a *non-classical* basis the reconstitutability of classical mathematics by Constructivism (which has already taken place) becomes more than a mere idle curiosity. Instead, it represents evidence that the non-classical schema is indeed the true logical basis of mathematics, just as it is of physics. Thus Frege was right all along; he was simply making use of too restricted a concept of logic.

The problem of class existence/inexistence which lies at the root of the undecidability of general set theory therefore requires resituating in terms of non-classical logic in order to achieve complete logical consistency, a consistency which involves the acceptance rather than the avoidance of undecidability. Furthermore, this undecidability at the core of mathematics is a profound testimony to the truth of the relativistic interpretation of mathematical objects supplied earlier. It is because of this interpretation that we can say that undecidability does not equate to irrationality and so can be embraced as a logical necessity within a fully rational but *post-classical* scheme, one which leaves philosophy, as it were, no longer epistemologically adrift as at the present time. Logical self-consistency is also restored to the foundations of mathematics, notwithstanding the apparent inexistence of classes.

The Cantor-Russell set paradoxes and the associated failure of set theory, highlight the fallacy of the notion of substantial existents (in this case classes) and hence the correctness of the alternative relativistic interpretation of universals. This is because the basic principle of set theory (the so called "class existence principle") is shown to be untenable by the paradoxes. It also highlights the error in the argument for so called "ontological commitment". Mathematics may in some sense be "indispensable", but its fundamental posits (i.e. the classes of general set theory) are revealed by the antinomies to be

---

[103] Of course we have to make the distinction mentioned earlier between problems which have been *proven* to be undecidable, such as the continuum hypothesis and those, like Goldbach's conjecture which are merely suspected (in some quarters) of being inherently insoluble. Of the former (which I call *objectively* undecidable) it maybe that their true status should be that of an axiom since, being objectively undecidable, they are in that respect axiom-like.

profoundly insubstantial. Numbers do not exist, it seems, in any *intrinsic* sense, instead they remain apriori-like due to the phenomenon of universal relativity.

This leaves only the issue of proof unaccounted for. Whilst in *classical* mathematics the concept of proof has an *absolute* character, in Constructive or non-classical mathematics it has a self consciously *relative* character instead. This I believe is a rational consequence of the implicitly non-classical basis of Constructive mathematics. Since the axiom systems of classical mathematics are not fully groundable on a *classical* basis (as Gödel has shown) then it follows that the *Constructivist* view of proof (which is implicitly relativistic) must be the valid one.

61. Gödel's Uncertainty Principle.

The Zermelo-Fraenkel system of axioms is consistent but incomplete (i.e. true statements are to be found outside it). If it were complete, however, it would be inconsistent. This trade off in the *decidable* foundations of mathematics (discovered by Gödel of course) is an analytical equivalent to Heisenberg's uncertainty principle. Both discoveries, on my interpretation, point to the universal presence of relativity in ontological and (ipso-facto) epistemological foundations alike. This indeed is their ultimate significance to wider philosophy.

The same point is also true of Popper's principle of falsifiability which is, indeed, merely a sub-instance of the more *general* principle of incompleteness. Indeed Popper's principle represents a remarkable bridge between the two worlds of the empirical and the analytical. One might say that Popper's principle is true because Gödel's *more general* theorem is true. But conversely we can say that Gödel's theorem is inevitable because the *ontological* truth implied by Heisenberg's principle (i.e. ontological indeterminacy) is the way *it* is. In other words, Gödel's result (though analytic) is a consequence of Heisenberg's *synthetical* result. And so, by a process of logical deduction we can say that Popper's principle (and the limitations on knowledge it implies) is a logical consequence of quantum indeterminacy.

Incidentally it has already, I believe, been mooted by others that we attempt to understand quantum mechanics in a framework of non-classical logic (implicitly this is already the case). The argument just supplied indicates that this understanding must be extended to *analytical* knowledge as well. At any rate, such an extension would amount to a complete theory of knowledge on self-consistent logical foundations.

62. The basis of Rational Philosophy.

My view, contra Kant, Brouwer or Frege is that number and geometry are alike *wholly* analytical and so *apriori*.[104] Although geometry is quasi-empirical in the sense that the objects of geometry are observable, their construction (as witness the *Elements* of Euclid) may be achieved solely with reference to apriori axioms.

And this, in point of fact, is the first step towards proving the thesis (which would greatly simplify matters if it were true) that the synthetic is really a *complex* form of the analytic. The first stage (as just taken) is to demonstrate that *geometry* is not synthetic (apriori or otherwise) – as according to Kant – but is in truth wholly analytic. Kant's arguments against this thesis it seems to me are weak and unsupported by the facts. They stem from an over-ambition on the part of Kant vis-à-vis the propagation of his new category – *synthetic apriori,* a category which has the effect of bringing philosophy to a grinding halt.

He incidentally gives unnecessary ground to the empiricists, most of whom, including Hume, would have been comfortable in granting that geometry is indeed analytical and therefore apriori in nature. But Kant's new category, as I have argued, is only a half way house towards a new and complete synthesis which is far clearer (because ultimately based on the analyticity of geometry and non-classical logic) than Kant's initial approximation to it. In a very real sense we will have eliminated the need to posit the category *synthetic apriori* at all and so Kant's great system (at once an aid and a blockage) will have finally outlived is usefulness.

The second and final step, the truth of which I intend to demonstrate in due course, is to prove that what we conveniently (and quite correctly) distinguish as *synthetical* is, at bottom, wholly *geometrical* in nature and therefore, given our aforementioned step, wholly *analytical* (and hence apriori) as well. Having done this the resurrection of Rationalism (which I call *neo-Rationalism)* will be complete. And empiricism (or what I term *neo-empiricism* as largely redefined by Popper) will be seen to be not a separate philosophy at all but rather an *instrument,* an *organon* of neo-Rationalism.

And all will be set, so to speak, within a justifying context of non-classical logic (what I call *neo-logicism),* which forms, therefore, the foundation of neo-Rationalism and of its organon neo-empiricism.

Non-classical logic is, of course, ungrounded and recursively justified, thus rendering the overall system of knowledge fully justified, rational, entirely analytical and complete. And this is and always will be the only correct general structure for a fully rationalistic and complete philosophy. It is, as it were, what the great system builders of the Rationalist tradition, from Plato to Kant, Frege and the Positivists were always tending towards but which, due to insufficiency of information and classical or anti-empirical prejudices, they never quite arrived at.

The problem faced by the classical rationalists in particular (for example, Descartes or Leibniz) was that the geometry they had to rely on (Euclidean geometry) was insufficient for the task of describing all phenomena. Additionally they had a tendency to dismiss or else ignore the *complementary* empirical approach, an approach which, in the long run, has (I will argue) confirmed their fundamental thesis, which is that the universe is *entirely* explicable (in a reductive way) as a *geometrical* phenomenon, a thesis first propounded by Plato nearly two and a half millennia ago and which I believe is now (as I hope to show) demonstrable.

Empirical analysis is of use to rational philosophy in that it ultimately confirms the validity of this basic thesis. Hence the error of the Rationalists in austerely denying themselves access to this instrument. But this point aside, apriori, non-classical geometry and non-classical logic are (in principle) entirely sufficient to fully describe the universe, including the aposteriori discoveries of the empiricists.

---

[104] The existence of prime numbers with their apparently random distribution is however a niggling contradiction of this view. It seems that the continuum of numbers manifests both an aposteriori and an apriori element (see appendix).

63. Two types of Indeterminacy.

It has been a fundamental error of analytical philosophy to suppose that the analysis of the rules of inference (or of linguistic use) would somehow eliminate the category of indeterminacy, whose epistemological and ontological import is truly central.

This is indeed the common error of both genres of classical analytical philosophy, the linguistic and the logicistic. *Principia Logica* has hopefully shown why such an ambition must be in vain and thus why the problems of philosophy will not go away by following current methods. However my work has hopefully also shown how both forms of analysis may be made viable by taking into account the lessons of non-classical logic. Ideal languages may be made to incorporate additional truth values for instance, whilst the procedures of linguistic analysis may recognize that certain formulations are neither true nor false, neither valid nor invalid statements and so, instead of dismissing these as pseudo-problems (a la Carnap) or as meaningless (a la Wittgenstein) learn to accept them instead as examples of problems which are inherently indeterminate and hence logically undecidable in nature. And also, these types of problem may, far from being non-existent as according to the current orthodoxy, be unlimited in number and liable to provide new fodder for debate perpetually.

Indeed an exact analogy with mathematics is possible, given the discovery that irrational numbers, initially thought to be non-existent (by the Pythagoreans for example) and then assumed to be rare, infact far *outnumber* the rationals. Thus, as it were, (to prosecute our analogy with philosophical problems), indeterminate numbers far outnumber determinate in terms of cardinality. Similarly, indeterminate problems of philosophy (which by their inherent nature can never be solved or even wished away) far exceed determinate ones in a similar fashion. And indeed the annals of contemporary philosophy alone are full of such inherently undecidable problems, earnestly under academic discussion even as I write, thus indicating the failure of analytical philosophy in even indicating the general nature of the issue correctly let alone in disposing of it.

And, in order to complete our analogy, simply to dismiss these indeterminate questions as pseudiferous, infernal or meaningless (as though they were epistemologically inferior to other types of question) is precisely akin to dismissing irrational numbers as either invalid or else as somehow less important or inferior to the rationals. Determinate and indeterminate philosophical questions are thus akin to determinate and indeterminate numbers in mathematics, reminding us perhaps of Tarski's extension of Gödel's incompleteness proof to higher orders of (non-mathematical) language.

There is, nonetheless, a significant problem with altering the procedures of ordinary linguistic analysis so as to take account of indeterminacy (something I advocate), which is that we usually *do not know* whether a problem is *definitively* indeterminate or not. Again, the analogy is with mathematics in that certain mathematical problems, as I have already noted, lie in this category as well, for example Goldbach's conjecture (which might be called *prima-facie indeterminate*). Other mathematical problems however, such as the continuum hypothesis, have been *proven* to be inherently indeterminate however, an epistemological status which might sensibly be titled *objectively indeterminate* in order to distinguish it from the former case.

Most problems of philosophy fall into the former, somewhat weaker category (of prima-facie indeterminacy), which should therefore be seen as defining a distinctive truth value in its own right. In other words, the truth value *indeterminate* disaggregates, in all but the simplest of logical analyses, into at least the two separate truth values which we have just defined. And something of the power of non-classical logic is thus revealed in that it is able to treat of such valuable distinctions without also doing violence to or excluding the more traditional truth values of classical logic. In attempting to do without such an expanded logic, analytical philosophy has unnecessarily hampered itself and condemned itself to incompleteness and ultimate failure, a failure which is, I believe, quite unnecessary.

This analysis has also hopefully served to illustrate the relationship between philosophy, conceived of as a system (as in previous sections), and philosophy conceived of as a critical pursuit or activity. In effect, philosophy is a system first, one which is ultimately grounded in fully recursive non-classical logic and an activity second. In other words it can be characterized as both system *and* procedure. But establishing the true general form of either of these two aspects of a complete philosophy has proven the undoing of all traditional philosophy.

In truth, however, the latter (analytical) conception is incorporated into the former (synthetical) conception as an *instrument* of the overall system. And this instrument (i.e. the *act* of logical analysis) allows us to *transcend* the problems of philosophy (through efficient categorization of them) rather than by simply dismissing them (since they are infinite in number) as pseudiferous or even as meaningless.

64. The Trivalent Basis of Induction.

A scientific law (for instance Newton's second law of motion) is deduced entirely from a finite set of recorded instances. What distinguishes induction from deduction is merely the *inference* that the set of relations deduced from finite or local data is of general or universal significance. Except for this postulated inference (which can never be proved only refuted) induction is really nothing more than a glorified form of *deduction*, of deduction in disguise.

However, the tendency to *make* the inductive inference – which is simply the act of generalizing from particular (often singular) instances - would seem to be part of the innate structure of the human mind, the product of millions of years of human evolution. Indeed, the process of learning is itself essentially inductive in nature, even when it occurs unconsciously, as in childhood. Scientific methodology merely adds the crucial elements of precision and rigour to the inductive procedure.

Since inductive inferences are continuously being drawn it is perhaps meaningless to speak of "pure observation" as though observation were ever wholly independent of concept and prior theory. Thanks to Kant therefore we have learnt to question the naïve picture of induction as a theory independent activity and this point, which stems from Kant, has therefore been a major theme of the philosophy of science in, for example, the writings of Karl Popper and Thomas Kuhn. Nevertheless and contrary to the opinion of Karl Popper on the matter, it *is* the process of induction which leads to the accumulation of all synthetical knowledge. The fact that, as Popper has unwittingly demonstrated, induction rests on *trivalent* logical foundations does not negate its validity. The death of the principle of verification (about which Popper is absolutely correct) does not ipso-facto contradict the inductive procedure as well, although it does force us to reappraise it (with respect to its true logical foundations). The validity of the inductive inference is *not* dependent on the validity of verificationism and it is Popper's greatest mistake to suppose that it is.

65. Neo-Foundationalism.

Carnap's supposition that "philosophy is the logic of science"[105], although it has been abandoned by the mainstream analytic community, is nevertheless largely true, a fact demonstrated by Popper's analysis of induction. Carnap's view (the basis of early logical empiricism) was abandoned because the question of *which* logic underpins the sciences (and we should extend this question to include mathematics as well, thereby making Carnap's statement more accurate) was more complex than Carnap or the logical positivists could reasonably have expected. Indeed, in *Meaning and Necessity* Carnap seems to have abandoned this view himself, expressing instead his own distinctive position on ordinary language analysis.

Notwithstanding his complete clarification of the problem of induction (something Kant, Carnap and Quine all ran aground on) even Karl Popper has failed to identify the true logical basis of the inductive procedure, i.e. that it is not, as Carnap and Quine had supposed, dependent on bivalency. Induction and deduction, though they are logical *procedures* (as mentioned before) are not *themselves* logic, (we cannot stress this distinction enough). Consequently, the logical basis supplies their justification and is therefore *prior* to them and so it is important (as Popper failed to do) to identify what this basis is. And this logical basis is non-classical and in particular trivalent. Popper thus missed an opportunity to cement himself at the forefront of formal logical philosophy where he undoubtedly belonged.

It is therefore, as already stated, the fundamental thesis of *Principia Logica* that non-classical logic forms the epistemological basis of all human knowledge and, in particular, of the sciences and mathematics. Indeed, Gödel's use of modern logic to investigate and prove the incompleteness of mathematics vis-à-vis its classical assumptions, demonstrates another basic thesis of this work which is that logic is not only prior to the sciences but prior to mathematics as well, a suspicion which seems to date back to Aristotle, but which, as I say, has quite recently been abandoned.

So what this interpretation amounts to is nothing short of a resurrection of *Foundationalism,* albeit on a new, non-classical basis. Furthermore, it is a third major thesis of *Principia Logica* that even criticism of non-classical logic presupposes its truth. In effect, scepticism, as a philosophical position, *itself* has an inescapable logic, and that logic is non-classical. Thus the discrediting of classical foundationalism (of foundationalism based on bivalent assumptions) is obviated.

This, which I call *neo-Foundationalism* is thus the only possible basis for a logical philosophy, one which instantiates relativism at its very core instead of attempting to stand in opposition to it, which was the fundamental mistake of classical foundationalism. For only non-classical logic (which is the correct form of logic) is capable of transcending the fundamental opposition between relativism and foundationalism, thereby presenting us with the comprehensive benefits of both perspectives; precision and formality (in the case of Foundationalism) and universality (in the case of Relativism).

---

[105] Rudolf Carnap, *On the Character of Philosophical Problems,* Reprinted in *The Linguistic Turn,* Ed R.Rorty, University of Chicago Press, Chicago, 1992.

66. Induction and Revolution.

What is perhaps the most striking thing about induction is not that it leads to the accumulation of our synthetical knowledge but that it is also responsible for revolutions in theoretical structures. I believe that it is precisely *because* induction leads to the accumulation of banks of knowledge that it is *also* able to trigger revolutions in the theoretical base (after all, observational anomalies *can* only occur relative to prior theoretical expectations). Transformations occur when what might be called points of *phase transition* in knowledge itself are arrived at.

The catalyst for transition is when a disjuncture occurs between accumulated background knowledge and new observations. On those rare occasions when new observations do not fit into existing theoretical structures it is likely, if the anomaly is indeed significant, that a more or less radical transformation of those structures will be attempted. Either adjustments will be made, thereby refining elements of the old theory, or else, if the anomaly is sufficiently challenging, entirely new theoretical structures will have to be constructed in order to incorporate and account for the anomaly. In that case the old theory will commonly be retained as a *special case* of the new and more general theory, thereby maintaining the continuity of science.

A third possibility, which may very well be the most common of all, is that theories continue to survive unaltered in the face of teething troubles or even fundamental anomalies, due to the absence of more compelling alternatives. Given this fact it is indeed correct to criticize Popper's view of falsification (sometimes called *naïve falsificationism*) as too simplistic. Scientists can only be expected to abandon accepted theoretical structures when new and proven ones of a more complete nature become available to take their place. Until such a time, as Kuhn has pointed out, scientists will always live with the old theory, no matter how many anomalies it harbours. Nevertheless Popper is correct in that it is the discovery of anomalies which generates the process of new theory formation, be it in never so roundabout a fashion.

So the process of change is an entirely natural one even though it lends the development of science a superficially radical or revolutionary outward appearance. As Popper has observed;

> "It is necessary for us to see that of the two main ways in which we may explain the growth of science, one is rather unimportant and the other is important. The first explains science by the accumulation of knowledge; it is like a growing library (or a museum). As more and more books accumulate, so more and more knowledge accumulates. The other explains it by criticism: it grows by a method more revolutionary than accumulation – a method which destroys, changes and alters the whole thing, including its most important instrument, the language in which our myths and theories are formulated… There is much less accumulation of knowledge in science than there is revolutionary changing of knowledge. It is a strange point and a very interesting point, because one might at first sight believe that for the accumulative growth of knowledge tradition would be very important, and that for the revolutionary development tradition would be less important. But it is exactly the other way around."[106]

Popper is thus the first person to have emphasized not only the revolutionary character of scientific development, but also its complementary dependence on *tradition,* a dependence which tends to be under-stressed by other philosophers of science.

---

[106] Karl Popper, *Towards a Rational Theory of Tradition*. 1948. Reprinted in *Conjectures and Refutations*. P129. Routledge, London, 1963.

Notwithstanding Popper's priority however it is to Thomas Kuhn of course that we owe the most detailed and pains-taking analysis of the inherency of radical theoretical change in the development of science.

Nevertheless, it is not the *novelty* of hypotheses (which, by their very nature, will *always* be striking, even revolutionary in appearance) but rather their *testability* which is important and which will always (except in those cases where facts are sparse) act as a vital restraint on novelty for novelty's sake. Popper's analysis of the logic of induction is thus a more fundamental account of what makes science effective (notwithstanding the above criticism of "naivety") than Kuhn's *complementary* account (not merely endorsed but preempted by Popper) of changes in the theoretical super-structure, although both are important to our overall picture of scientific development.

To borrow an analogy from Marxism, induction is the *base* or driver of scientific development whereas our theories, hypotheses and conjectures are the *superstructure*. The base (the primary object of Popper's analysis), as the driver of scientific development, never changes and is rooted in timeless non-classical logic, whereas the superstructure (the primary object of Kuhn's analysis) is subject to often radical transformation as a result of changing knowledge conditions. It is therefore *not* timeless and unchanging, although, when knowledge approaches completeness the superstructure may (in contradiction to Popper and Kuhn's views on the matter) approach stability.

The analogy can thus also be made to the evolution of species. The outward form of species is constantly changing, but the *mechanisms* of genetic change are comparatively timeless. Nevertheless, when a species enters into equilibrium with its environment the underlying mechanisms that commonly generate change can also (and perhaps rather more commonly) enforce stability in the phenotypical superstructures.[107]

Given the secure base to the objective development of science provided by trivalent induction the perennial radical transformation of the theoretical superstructures of the sciences should not be the cause for dismay and nihilism that it is often made to appear by critics of science such as Paul Feyerabend. There is, at any rate, nothing anomalous in it. Furthermore, radical transformations in the superstructure are likely to become less frequent as information becomes more complete (due to induction). That is, as our theories become less *underdetermined* so they become more stable and unlikely to change.

This is because (as we remember from Moore's paradox of analysis) complete knowledge of an object places an absolute break on the formation of new empirically significant hypotheses about that object, simply because there is no new information to make such hypotheses necessary. Consequently, the continuous paradigm shifts prophesied by Kuhn will become increasingly impossible as (and if) information approaches completeness. And this is so notwithstanding the fact that, largely for technical reasons, completeness of information is unlikely ever to be reached. After all, many autonomous parts of science (for example chemistry) have already achieved completeness and therefore meaningful paradigm shifts concerning their representation are most unlikely ever to occur.

---

[107] For a more detailed discussion concerning the inter-relation of issues surrounding natural and cultural selection see part two of this work.

67. The Rendering of Classical Philosophy.

Thus, when hypotheses are anchored to the inductive method, with its basis in three valued logic, they can never be *entirely* arbitrary. And this unique stricture indeed is the secret of the epistemological primacy of the scientific method, notwithstanding epistemological and ontological relativity. For what distinguishes science in contrast to any other epistemology (for example Quine's Homeric gods or Feyerabend's equally dogmatic witchcraft) is the *anti-dogmatic* principle of testability.

Since induction is founded in a logic of three rather than two values as hitherto thought it does not, as traditionally assumed, contradict the notion of epistemological and ontological relativity as many critics of science have assumed. This is because induction, based on three values, does not presuppose the truth of its inferences, as previously thought, which are instead subject to perpetual empirical examination and (technically speaking) indeterminacy. And since what is called "common sense" or "folk" knowledge is entirely based on a less rigorous form of inductive inference it therefore follows that nothing of value or truth is excluded from a system of knowledge which is based upon it.

Furthermore, the work of Popper, Kuhn, Lakatos and others have clearly established that the underlying logic of discovery in not only science but in mathematics as well is *inductive* (and hence also deductive) in nature and this we may assume is due to the applicability of non-classical foundations to these foundational subjects upon which *all* our knowledge effectively depends.[108]

And, given that critical or sceptical philosophy may in principle construct or explode *any* conceivable concept, distinction or position its ability to do this relies on epistemological and ontological indeterminacy and hence on universal relativity. The *logical* basis of such a comprehensive philosophy is thus non-classical in nature and so both sceptical and critical philosophy presuppose non-classical logical foundations of the sort described in *Principia Logica*. It is not therefore feasible to utilize scepticism as an alternative position to the one outlined here. The error of seeing sceptical and constructive or logical philosophy as fundamentally opposed is thus also (I think for the first time) transcended.

Indeed the traditional lack of consensus in philosophy, normally regarded as indicative of the fundamental bankruptcy of philosophy should instead be seen as further circumstantial evidence concerning the validity of universal relativity and the non-classical epistemological basis discussed earlier in this work. Other than this the traditional problems of philosophy have (except for the sea of inherently indeterminate, para-linguistic problems) mostly devolved to the sciences themselves, notably to physics, psychology and linguistics. Furthermore, even the question of the *epistemological basis* could fairly be interpreted as a purely *formal* question belonging to the domain of pure (non-classical) logic, thereby completing the much needed stripping down and rendering of classical philosophy.

Logic, being apriori, is not subsumable to subjectivity, which point is the fundamental criticism (in my view) of *Phenomenology* and *Replacement Naturalism* alike. Philosophy is therefore nothing if it is not the study of non-classical logic and its numerous fields of application. Certainly, in as much as philosophy is anything at all (and if we say that formal logic and its applications are distinct from philosophy then we effectively deny philosophy any substance at all) then it is this and this alone.

---

[108] See Lakatos, Imre, *Proofs and Refutations,* Cambridge University Press, Ed J. Worrall and E. Zahar. 1976.
Lakatos in many ways imports the discoveries of Popper and Kuhn to account for the more rarified world of mathematical discovery. Though this is I think his principle achievement he also pays ample dues to Euler and Seidel as important antecedents who also detected the role of inductive thinking in the framing of mathematical theorems. Although Lakatos deprecates the term induction somewhat it is nevertheless clear that his theories and those of his antecedents pertain to the centrality of the inductive as well as the deductive inference in the framing of mathematical theorems.

68. The Trivalent Foundations of Mathematics.

One of the most interesting implications of the incompleteness theorem is that, given the uncertainty surrounding the axiomatic base, mathematics would seem to be placed on a similar epistemological footing to the exact sciences. Indeed, Gödel draws this disturbing conclusion himself;

> "In truth, however, mathematics becomes in this way an empirical science. For if I somehow prove from the arbitrarily postulated axioms that every natural number is the sum of four squares, it does not at all follow with certainty that I will never find a counter-example to this theorem, for my axioms could after all be inconsistent, and I can at most say that it follows with a certain probability, because in spite of many deductions no contradiction has so far been discovered."[109]

That is, given the complex interconnection of proofs and the uncertainty of the axioms it is impossible to rule out the possibility that a counterexample may not enter into the system and upset whole chains of proof, like a computer virus. This effectively means that, after Gödel, mathematical theory is, in principle, subject to the same kind of falsification as scientific theories are. This further suggests that the epistemological basis of mathematics and the sciences (which we have identified as non-classical logic) are, at bottom, identical.

All of which implies a *Popperian* solution to the problem of the axiomatic base, which is that the base remains valid only for as long as a counter-example is not found against it (as distinct from a merely undecidable statement). In this case the epistemological status of an axiomatic system (such as ZFC for example) would be precisely that of a physical theory (such as quantum mechanics) which only holds validity as long as some counter instance is not provided against it. Thus Gödel's result means that the principle of falsifiability *applies to axiom systems just as it does to scientific theories.* This indeed confirms that the logical basis of mathematics, like that of physics, is ultimately trivalent. And this observation represents the objective end of epistemology since it places knowledge – *synthetic and analytic alike* – on objective foundations notwithstanding indeterminism. It also suggests that, as a discrete philosophy in its own right, Empiricism, like mathematics, is incomplete with respect to Constructive Logic.

For Gödel, as for post-analytical philosophy this definitive solution was not apparent however (no one had thought of non-classical foundations in this context or of the principle of falsification as the correct application of this logic to the foundation problem) and these problems, (prior to *Principia Empirica* that is,) still remain in the same form as that faced by Gödel.

---

[109] Kurt Gödel, *The modern development of the foundations of mathematics in the light of philosophy* (Lecture. 1961) Contained in; Kurt Gödel, *Collected Works*, Volume III (1961). Oxford University Press, 1981.

69. The Foundations of the Apriori.

So what is Gödel's suggested solution to the general problem of epistemology which includes the problems of induction and incompleteness as special cases?

Gödel (op cit.) recommends Kantian idealism and in particular Husserl's Phenomenology as the best solution to the problem of epistemology. This is presumably because these arose specifically with a view to solving the problem of induction which, it turns out, is very similar to the problem of incompleteness underlying mathematics.

As Gödel correctly observes, his theorems reduce analytical knowledge to the same epistemological status as indeterminate *empirical* knowledge. Thus the solution to the problem this poses is more or less *identical* to that of the problem of induction, the solution being (as we have now identified it) a turn to constructive and in particular to three-valued logic, a logic which is not dependent on an axiomatic base and so is not subject either to the problem of incompleteness or of inconsistency.

In effect this and not Idealist phenomenology is the correct formal solution to the problems of epistemology raised by both Gödel and Hume alike. It re-establishes the apriori basis of logic (which was undermined by Church's theorem as it applies to classical logic) and hence of mathematics and inductive science as analytical exercises. Indeed, logic and mathematics can only be considered apriori activities if their basis is held to be non-classical. And the reason why classical logic is found to be incomplete and hence *not apriori* is because classical logic was always an artificial basis for logic since the law of excluded middle is clearly an arbitrary restriction.

Once this restriction is taken away logic is seen in its true, fully recursive, complete and apriori guise and so epistemology based on solid foundations (truly apriori foundations) becomes possible again, thereby incidentally rendering Kant's unwieldy and undecidable system of categories and intuitions somewhat redundant and unnecessary. And this surely is the greatest breakthrough in formalizing epistemology on objective, truly apriori foundations.

In contrast Kantianism and Phenomenology represent an incorrect and informal solution to what amounts to a problem of logic. Infact they represent a gigantic *category mistake*, the largest in the history of philosophy, *since solving a problem of logic using classical metaphysics is inherently absurd.*

As temporary solutions the systems of the Idealists were valuable and even (approximately) correct. But it is now clear that the problems are infact soluble with a high degree of precision with recourse to the methods of non-classical logic, and this is surely a decisive improvement. If determinate knowledge can be retained as a special or contingent case (which it can, on a constructive basis, as Popper has perhaps unwittingly proven) then this should be sufficient for us, especially as it solves *all* the problems specifically addressed by Kant.

An additional problem for Gödel's appeal to Phenomenology as a means of obviating nihilism is that Heidegger, working from within the same Kant-Schopenhauer-Husserl tradition which Gödel appeals to deduces Ontological Nihilism from this tradition.[110] Thus, even if we accept the epistemological priority of Kantianism and phenomenology over that of apriori non-clasical logic (which is a mistake made by the Intuitionists themselves, ironically) then the nihilistic implications of incompleteness are still not escapable as Gödel had hoped.

But, in my view, the fundamental problem posed by Idealist epistemology (of whatever form) in contrast to that offered by apriori logic, is the problem of *precision*. Since their understanding of the problem of epistemology lacks precision so too does the solution they offer and this is (I believe) *entirely* explicable as a consequence of operating in an era or milieu which predates the advent of modern formal logic. Other than this they, and Kant in particular, were pointing in the right direction, particularly in their use of the concept of apriori sources of knowledge, sources of knowledge, that is, capable of shoring up the foundations of the empirical sciences.

---

[110] (see Martin Heidegger, *Was ist das Metaphysics,* 1926).

Furthermore, the failure of modern logic to solve the fundamental problem of epistemology (a failure which has analytical philosophers running to Phenomenology for help) is also entirely explicable as being a consequence of an erroneous understanding of the true basis of apriori logic which is, of course, non-classical.

70. The Limits of Phenomenology.

But to illustrate our thesis of the imprecision of idealism let us examine the solution to the fundamental problems of epistemology offered by perhaps the most highly respected post-Kantian system; Husserl's Phenomenology. Of the problems of epistemology Husserl has this to say;

> In fact, if we do not count the negativistic, sceptical philosophy of a Hume, the Kantian system is the first attempt, and one carried out with impressive scientific seriousness, at a truly universal transcendental philosophy meant to be a *rigorous science* in a sense of scientific rigour which has only now been discovered and which is the only genuine sense.
> Something similar holds, we can say in advance, for the great continuations and revisions of Kantian transcendentalism in the great systems of German Idealism. They all share the basic conviction that the objective sciences (no matter how much they, and particularly the exact sciences, may consider themselves, in virtue of their obvious theoretical and practical accomplishments, to be in possession of the only true method and to be treasure houses of ultimate truths) are not seriously sciences at all, not cognitions ultimately grounded, i.e., not ultimately, theoretically responsible for themselves - and that they are not, then, cognitions of what exists in ultimate truth. This can be accomplished according to German Idealism only by a transcendental-subjective method and, carried through as a system, transcendental philosophy. As was already the case with Kant, the opinion is not that the self-evidence of the positive-scientific method is an illusion and its accomplishment an illusory accomplishment but rather that this self-evidence is itself a problem; that the objective-scientific method rests upon a never questioned, deeply concealed subjective ground whose philosophical elucidation will for the first time reveal the true meaning of the accomplishments of positive science and, correlatively, the true ontic meaning of the objective world - precisely as a transcendental-subjective meaning.[111]

As we have already seen, the epistemological validity of the natural sciences is not dependent on "self-evidence" or on a "deeply concealed subjective ground" as Husserl assumes, but rather more mundanely on the trivalent logic described by Popper in the *Logik der Forschung*. Popper's discovery predates Husserl's writing here by a number of years and had Husserl been more familiar with it and had he understood its implications (something which seems to be a general problem amongst philosophers) then his sense of the crisis of the so called "European sciences" (essentially a crisis, triggered by Hume, concerning their epistemological basis) would have been considerably eased. Kant's verbose riposte to Hume's scepticism at least has the excuse of being pre-Popperian, but this is not an excuse available to Husserl and his repetition of Kant's idealism, or to Quine or Carnap or to other still more contemporary philosophers for that matter.

It is, in any case, difficult to see how idealist philosophy could ever be a "rigorous science" in any meaningful sense. The only means by which philosophy can hope to approach the rigours of modern science is through the adoption and application of the truly ground-breaking discoveries of modern and post-classical logic, discoveries which alone offer the possibility of a sound and precise basis for epistemology as I have hopefully shown. And I have hopefully also demonstrated why despair in this (foundationalist) project is premature and hence why recourse to psychology, Phenomenology or Kantianism (the only viable alternatives perhaps) is also premature.

At any rate, we agree with Husserl that the empirical sciences are not self guaranteeing and so are not "theoretically responsible for themselves". We disagree with Husserl (and Quine) in that we see that the application of non-classical logic and not Phenomenology or psychology is the correct formal and alone

---

[111] Edmund Husserl, *The Crisis of European Sciences and Transcendental Phenomenology* (1954) publ. Northwestern University Press, Evanston, 1970. 1937. Section 27.

precise solution to this problem. Similarly Kant's solution (the categories and forms of intuition as alternative sources of apriori support), though it represents the best early approximation to the views developed here, is superfluous since apriori foundations are supplied by logic alone thus rendering Kant's indeterminate categories unnecessary.

Finally we repudiate the solution of the logical positivists since although their systems are founded, correctly, in apriori and precise logic, it is an incorrect and incomplete conception of logic, one which is unnecessarily weighed down and distorted (beyond effectiveness) by the acceptance of the wholly artificial law of excluded middle (tertium non datur) of classical logic. As a result of this structural flaw and as Quine and every other commentator has pointed out, the foundationalist project of classical Logicism and logical empiricism, has demonstrably failed.

The reason we have identified for Popper's success in solving part of the problem of epistemology (that part which pertains to the problem of induction) lies, as I have said, in his unwitting application of non-classical logic to the problem. As I have described in section sixty one, precisely the same solution can be applied to solving the problem of epistemology as it applies to analytical foundations – i.e. the foundations of mathematics. These two problems solved – the problem of the foundations of *Synthetical* knowledge and the problem of the foundations of *Analytical* knowledge – it is entirely fair to say that the problem of epistemology (the last genuine problem left to philosophy) is completely resolved on a precise logical basis.

71. What is Metaphysics?

It is thus the fundamental thesis of this work that the epistemological foundations of the sciences *is indeed* metaphysical (since, as Husserl has noted, the sciences are not self validating), but that these metaphysical foundations are identical with apriori post-classical logic. Thus, since the true basis of metaphysics was all along nothing more than this it is possible to argue that metaphysics has indeed been vindicated but that it has also been established on an unwavering and complete formal basis.

Metaphysics thus reaches its conclusion *only* in and as logic and its applications; it is and can be no more than this. All other possible bases for it, from Plato to Kant and Husserl can therefore be dismissed as worthy but ultimately inadequate alternatives, approximations or tendencies toward this. Similarly the positivist and empiricist denigration of metaphysics can also be dismissed since it is clear that physics (or psychology) does not account for the epistemological basis of the sciences. Only *logic*, which is suitably apriori and so infinitely resilient, can do this. Metaphysics is thus redeemed (by placing it on a purely formal basis) rather than banished (as Carnap and Wittgenstein had hoped to do). And it is apparent, as Kant intuited, that empirical knowledge is deeply interfused with and bounded by its apriori character.

Logic, being apriori, is not reducible to subjectivity, which point is the fundamental criticism (in my view) of *Phenomenology* and *Replacement Naturalism* alike. Philosophy is therefore nothing if it is not the study of non-classical logic and its numerous fields of application. Furthermore, this study renders Kant's search for other sources of the apriori unnecessary. Certainly, in as much as philosophy is anything at all (and if we say that formal logic and its applications are distinct from philosophy then we effectively deny philosophy any substance at all) then it is this and this alone.

The question of whether the world is "ideal" or "real" is thus neatly obviated as being at once undecidable and irrelevant and a new and objective basis for metaphysics is finally established in which this question need not arise. The only counter to this is to say that logic is somehow subjective. But few would doubt its apriori status, a status which renders the question of the subject, (such a central concern for Phenomenologists,) ultimately irrelevant.

After all, the purpose of Kant's system had been to obviate the problem of subjectivity (first identified by Descartes) by finding a locus for knowledge which is apriori – hence the postulation of the categories and the forms of intuition. We have achieved the same end in a far more minimalistic way, relying only on the aprioricity of (non-classical) logic itself. And it is surely no coincidence that not only is logic uniquely apriori (from an objective point of view) but it is also *prior* to the sciences and mathematics in the ways we have described, suggesting its status as the true foundation of knowledge which Kant had nobly searched for and which the classical Logicists had come so close to describing and which the post analytical philosophers have so ostentatiously abandoned.

Of course, plenty of things are *debatably* (or undecidably) apriori (Kant's categories being only the most pertinent example) but only (non-classical) logic, being fully recursive, is *truly* and indisputably so. And, from the strictest viewpoint, as Gödel has shown us, even mathematics, stripped of its logicist foundations, cannot be considered truly apriori. Hence *neo*-Logicism must form the basis of an objective and fully *metaphysical* account of the foundations of rational human knowledge. Nothing else will do and nothing else (remarkably) is needed.

72. The Problem of Ontology.

Looked at closely then, Phenomenology turns out to be little more than another failed attempt to solve the problem of induction, a problem which constitutes one half of the fundamental problem of epistemology (the half that pertains to *synthetical* knowledge that is). We may additionally observe that Phenomenology is a variant of Kantianism just as Replacement Naturalism is a variant of Phenomenology.

Nevertheless Husserl makes another valid observation (again reminiscent of Kant) when he argues, in effect, that the solution to the problem of epistemology will also supply the solution to the problem of ontology as well (what Husserl portentously calls "the true *ontic meaning* of the objective world")[112]. This we have already found to be the case since both problems have their objective foundations in the category of indeterminacy. And this should not be surprising since the problem of epistemology is necessarily a special case of the wider problem of ontology. To put it another way, knowledge becomes a problem as a by product of the fact that (as has been established) the putative *object* of knowledge is without determinate foundation or substantial presence.

In contrast to *this* difficulty the difficulties surrounding the *subject* as locus of knowledge, which is the main focus for all forms of Phenomenology, (and which is solved with reference to the aprioricity of logical analysis) is, I think, comparatively minor.

However, although the expansion and correction of logic (by means of the abandonment of tertium non datur) restores rationality to the epistemological base and to the superstructure of the sciences alike, it only does so at a price. And this price is the loss of classical certainty and the concomitant acceptance of universal relativity and indeterminism, albeit on an objective footing supplied by logic and physics. In a sense even certainty remains, but it is the non-classical certainty that indeterminism is both universal and inherent.

We should at any rate realize that our goal must be to provide a logical account of the grounds for the possibility of contingent knowledge and not, as the Phenomenologist supposes the grounds for certainty or for some non-existent "ultimate truth". This granted (and I take the case to be overwhelming) then epistemological foundations can be deemed to be secure, without any terminological imprecision.

---

[112] Husserl, loc. cit.

73. The Ontological Deduction.

Through the so called "transcendental deduction" Kant deduces objectivity as a logical consequence of the self-evident or axiomatic existence of a Cartesian subject. He thereby completes a major aspect of the Rationalist project as revived by Descartes. Descartes, we may remember makes what might be called the "subjective deduction" ("Cogito ergo sum") from which, apparently, we may at least be confident of our own subjective existence if of nothing else. Kant extends this insight by observing that, without an external world, at once transcending and defining the subject, then the subject would have no definition and hence no identity.[113] Thus Descartes' subjective deduction *ipso-facto* entails a transcendental deduction which extends the argument for ontological reality to the inter-subjective domain as well. To put it another way; subjectivity (whose axiomatic character is demonstrated by Descartes) necessarily entails objectivity (or inter-subjectivity) as a logical corollary.

From this basis Kant is then able to analyze the apriori constituents of empirical experience whose apriori character is now guaranteed through reference to the transcendental deduction. This analysis (of the "categories" and the "forms of intuition") represents the heart of Kant's noble system, whose implications (for objectivity) are then extended to an analysis of ethics, aesthetics and axiology in general. And yet the foundations of this entire system rest on the validity of the transcendental deduction and on Descartes' earlier subjective deduction. Through this system Kant is able to effectively explode the simplistic distinction between Idealism and Realism, a theme subsequently amplified by Husserlian Phenomenology.[114]

As we have seen, intersubjectivity is facilitated by the apriori principle of symmetry, itself a logical corollary of the parsimony of the physis. Indeed, were it not for the counteracting force of the *first* engineering function of the universe (forbidding states of either zero or infinite energy) then the principle of symmetry (what I call the *second* engineering function of the universe) would simply translate to the existence of nothing at all.

Since there is apriori order (i.e. net conservation of symmetry) then inter-subjectivity becomes possible, even inevitable. This, *logico-empirical* analysis complements well Kant's transcendental deduction and so serves to account for what Kant identifies as the fundamental mystery of epistemology; "Cognition apriori as to the objects of experience."[115]

Nevertheless it can be demonstrated that Kant's transcendental deduction contains some unexpected consequences for his system. For it follows from the transcendental deduction that all entities – including the Cartesian self – are ultimately defined by and therefore ontologically dependent upon *other* beings and entities that are of an *extrinsic* nature. Thus the transcendental deduction may easily be flipped around so as to confirm what might be called the *Ontological Deduction* – which is that all posited beings lack intrinsic existence. Furthermore, as we shall see, Kant's failure to make this deduction (a deduction only ever arrived at in Buddhist philosophy) leads to unfortunate consequences that dominate philosophy after Kant.

---

[113] This is reminiscent of the founding idea of linguistic (Saussurean) structuralism – that words are defined not by something inherent but rather by the various functions they perform as part of an informal system of relations. This powerful relativistic idea from the era of Einstein retains (I think) its validity and is an aspect of structuralism and semiotics which survives modern critiques concerning the rigidity of structuralism.

In view of the remarkable parallels we have already hinted at between Buddhist and modern philosophy consider also the well documented foreshadowing of structural linguistics in the grammar of Panini who is commonly assumed to have flourished in the same era and milieu as the Buddha.

[114] It must again be pointed out, however, that this system, for all its ingenuity, does not solve the problem it initially set out to solve – Hume's problem of induction. This, it transpires, is a problem of *logic* and is eventually solved, with comparatively little fanfare, by Popper.

[115] Kant, *Prolegomena*. Loc. cit.

Incidentally, this relativistic implication of the transcendental deduction, never made explicit by Kant himself, may in part be interpreted as an extension of similar ideas in Leibniz's *Monadology,* where relativism is notably more explicit;

> "The interconnection or accommodation of all created things to each other and each to all the others, brings it about that each simple substance has relations that express all the others and consequently, that each simple substance is a perpetual living mirror of the universe."[116]

Leibniz therefore may be said to echo the (Buddhist) doctrine of universal relativity except for the fact that he maintains the ancient fiction (dating back to Democritus) of elemental and eternal "simple substances" – albeit (perhaps under the influence of Descartes and Spinoza) he identifies these elemental entities or "Monads" not with atoms (as Democritus and the materialists had done) but rather with souls or conscious entities. The being of the Monads (though *everything else* in Leibniz's system has a relative existence) is ultimately *intrinsic* and therefore not subject to the troublesome implications of relativism. Thus, rather inconsistently, *Monadology* combines a vision of logical relativism with one of metaphysical essentialism.

In Kant however this commitment to Aristotelian essentialism is *even more* eroded (due, no doubt, to the influence of Hume) and may be said to have been relegated to the tenuous notion of a so called "ding-ans-sich" or (more pompously) "noumenon". But given the relativistic implications of the transcendental deduction even this tenuous posit is unsustainable, a fact never quite grasped by Kant.

Indeed, this lacuna on the part of Kant may be said to be responsible for the errors at the heart of the subsequent phase of German philosophy after Kant. In the case of Hegel for example the gaseous notion of a noumenon is drafted in as inspiration and support for the so called "Phenomenology of Spirit", whereas in the case of Schopenhauer, (Hegel's chief and ultimately successful rival to the heirship of Kant) the same gaseous and unsustainable notion is adopted as the central justification of Schopenhauer's immensely reified concepts of the "Will and Idea". And these latter two notions in turn give life to similar concepts in Nietzschian metaphysics and in Husserlian phenomenology respectively. Thus all post- Kantian philosophy of significance except for Analytical philosophy is to some degree or other in thrall to a demonstrable and highly regrettable error.

The whole Rationalist project, from Descartes through to Kant and Hegel may therefore be said to operate on an idealized or classical concept of the subject as a monadic or fundamental posit, a concept not granted by the empiricists (who, like the Buddhists, view the self as a composite or *conditioned* entity) and unwittingly undermined by Kant's own transcendental philosophy. Indeed Brentano's Intentionalist interpretation of psychology is more in keeping with the implications of the transcendental deduction than Kant's own theory of the transcendental unity of apperception, a theory which does nothing to alleviate the nihilistic implications of the transcendental deduction. And it is under the influence of Brentano's Intentionalism that Husserlian Phenomenology and ultimately Heideggerian and Derridean nihilism come into existence. Thus, in effect, notwithstanding Kant's failure to admit the full implications of his theories, modern philosophy arrives at a species of ontological nihilism in any case. And after all, it could do no other.

However one conceives it therefore, Idealism, though inevitable (as Berkeley, Kant and Husserl all make plain) is ultimately self annihilating, due, fundamentally, to the irresistible implications of universal relativity.

Conversely, Empiricism is unsustainable as a complete philosophy in its own right since it leads to inevitable contradictions and inconsistencies. As we have seen, for example, the concept of existence transcends experience since the mere *idea* of absolute inexistence conceals an implicit Godelian (or perhaps Tarskian) contradiction. If absolute nothingness "existed", then absolute nothingness would *itself* be a state of being.

Thus *true* non-being takes a *relative* form of the sort perhaps best exemplified by our universe. That absolute nothingness necessarily takes a *relative* form is a point first attributable to the Buddhist

---

[116] Leibniz, *Monadology.*

philosophers and notably to Nagarjuna, but it is possible to arrive at this conclusion through the application of logic and *independently* of any observations. And this latter fact serves as our chief counter-example to the fundamental premise of Empiricism; "No concept without experience."

Thus Kant is effectively confirmed in his suspicion of the limits of Cartesian rationalism and the inherent incompleteness of Empiricism alike. At least prior to the advent of polyvalent logic Kant's alternative system has (in my view) remained the best solution to the limitations revealed by Kant's analysis.

74. Neo-Rationalism.

Relying solely on apriori concepts it is possible to construct logic, mathematics and geometry; an absolutely crucial fact which is implicit in the doctrine of *Constructivism*. This, we should note, *is everything except the (aposteriori) sciences.* In effect, what is *omitted* from apriori understanding is nothing more than the determination of which aspects of apriori geometry happen to obtain in our particular universe. In order to ascertain this mundane and even trivial detail we require recourse to *observation* which is, by definition, aposteriori in nature. And the associated procedure of observation and extrapolation from data is the *sole* contribution added by Empiricism to Rational philosophy. Empiricism is therefore not inherently at odds with Rationalism but instead constitutes a specific and subordinate aspect of the overall system – in a manner of speaking Empiricism is simply *applied Rationalism.*

Consequently, it is absurd to think of Empiricism as the "first philosophy", but equally it does not follow (as Quine seems to think it does[117]) that because *Empiricism* is not the first philosophy that therefore there is and can be no such thing as a first philosophy.

Furthermore, it should be apparent that, in granting Empiricism a co-equal status with Rationalism Kant (who began his career as a classical Rationalist) surrendered too much ground to the empiricists. And this mistake was almost certainly an over-compensation for the fact that Classical or Cartesian Rationalism had unaccountably over-looked or even denigrated the vital role of Empiricism which, as I have argued, should rightly be seen as an integral (but ultimately subordinate) aspect of Rational philosophy.

For, as I have argued, empirical analysis is simply the means by which the geometry of this particular universe is ascertained. It is nothing more than this and never does it (Empiricism) function independently of the apriori concepts of logic and geometry to which it is subordinate. And this latter point also sums up, in a crystal clear way, the (correct) critique of Empiricism implied by Kant's system and in particular Kant's imperfectly expressed concept of *synthetic apriori*. For Empiricism, (whether we are conscious of the fact or not), is little more than a *method*, the method of applied Rationalism.

Aposteriori analysis is thus not absolutely distinct from apriori analysis as has been assumed by modern philosophy but rather aposteriori analysis should be interpreted as representing a special or limiting case of apriori analysis, which is why I say that rational philosophy *precedes* Empiricism as the basis of all philosophy and is not merely coequal as Kant (under the influence of Hume's somewhat misguided attack) assumed.[118] This view I call *Neo-Rationalism.*

---

[117] Quine, *Epistemology Naturalized,* op. cit.
[118] My interpretation therefore is that Kant's system represents a good first approximation of the relationship between Empiricism and Rational philosophy but that it is nevertheless out of focus and lacking precision.

75. The Essence of Metaphysics.

The Empiricist attack on metaphysics, which began with Locke and Hume and which perhaps climaxes with the first phase of Analytical philosophy, could also be interpreted as part of the general assault on classical Rationalism. Although what I term neo-Rationalism places a special emphasis on apriori sources of knowledge (to wit logic and mathematics) which are by definition *meta-physical* it too deprecates the possibility of any other legitimate source of metaphysics. However, it is precisely this fact (that mathematics and logic *are* metaphysical) which points to the limits of the empiricist critique of knowledge and which accounts for something of the failure of analytical philosophy – at least, that is, in the first two phases of its development. After all, the basis for rational philosophy already laid down in this work – rooted in non-classical rather than classical logic – could reasonably be interpreted as the basis of a third and final phase in the development of analytical philosophy. For, analytical philosophy, whilst it has always felt the strong pull of empiricist assumptions has never itself been crudely empiricist.

Valid though the reaction was against what might be called the "bad metaphysics" of German Idealism (and in particular Hegelianism) it is apparent that the early analytical philosophers over-compensated in their anti-metaphysical zeal and, as a result, failed to grasp that the eradication of *all* metaphysics ipso-facto entailed the eradication of mathematics and logic (and hence the sciences) as well. This at any rate remained their well kept secret.

Furthermore, the essential emptiness of mathematics and logic – pointed out so triumphantly by the empiricists – has been readily granted and indeed elaborated upon in some detail by me as have various arguments (entirely missed by the empiricists) concerning the equivalent emptiness of physics and the subsidiary sciences (particularly in view of the strange elusiveness of *the given*). Consequently, the arguments *against* placing logic and mathematics (in that order) ahead of the sciences in terms of their generality are all found to be lacking.[119]

And the result of admitting this is the displacement of Empiricism as the dominant mode of modern philosophy by what I call neo-Rationalism. Coherency and (at last) the correct epistemological grounding of the sciences (along with a purged and objectified metaphysics) is the benefit derived by this approach. We might even say that the ambitions of the first phase of analytical philosophy, largely abandoned by the second phase are resurrected and entirely achieved in the third (whose elements are fully outlined in this work). Beyond this we have no need to look for a complete and fully recursive philosophical system resting on entirely objective foundations.

Since logic is apriori it cannot be dismissed as a transient phenomenon of the sort upon which the twin systems of Empiricism and Phenomenology (itself a species of empiricism) are ultimately built. Thus the charges of physicalism or psychologism do not apply to a system built on these (apriori) foundations.

Of course, since Gödel, Tarski and Church a fully *deterministic* logical philosophy has proven to be unsustainable, hence the collapse of the first phase of analytical philosophy (based as it was on a classical concept of logic and determinism). But with the *adjustment* of Aristotelian "logic" entailed by the abandonment of the law of excluded middle or *tertium non datur* the full power of logic, of logic in its true and proper form, becomes apparent. Early analytical philosophy can thus be said to have failed

---

[119] Mathematics, including geometry, is underpinned by logic since the ZFC system of axioms is ultimately a system of *logical* rather than mathematical rules. That which is omitted by ZFC is nevertheless captured by the polyvalent scheme of *non*-classical logic of which ZFC may be said to be merely a special or limiting case. Consequently, a neo-logicist schema may be said to be complete on account of the fact that it *incorporates* in a paradoxical, but nevertheless fully *objective* way, the twin concepts of incompleteness and indeterminacy.

After all, when Gödel exposed the pretensions of mathematics to classical completeness he did so by making use of *logical* arguments, as did Church when he exposed the analogous limitations of *classical* logic.

simply because its conception of what logic is was incomplete and deformed by an unsustainable Aristotelian assumption.[120]

Since mathematical proof always relies on arguments from logic it is apparent, I think, that Frege's intuition concerning the priority of logic was absolutely correct. It is simply the case that he, like his analytical successors laboured under an erroneous idea of what logic actually *is*. At any rate the fundamental thesis of this work is that if our view of logic is *corrected* (vis-à-vis the abandonment of Aristotle's manifestly artificial restriction) then the project of logicism and its extension to the sciences miraculously becomes possible once again and, furthermore, renders any *other* possible basis for philosophy (such as empiricism, scepticism or phenomenology) superfluous.

---

[120] What is of key importance is to understand that non-classical or constructive logic is not an *alternative* system of logic to classical (one with a few more convenient features) it is, far rather, the correct expression of what logic truly *is*.

76. The Roots of Philosophy.

Non-classical logic, which is the correct and undistorted form of logic, underpins general set theory and the axioms of mathematics which in turn underpin geometry and hence (as we shall argue) the whole of physics and the sciences as well. That which is not accounted for by classical mathematics (due to incompleteness) is nevertheless adequately incorporated by non-classical logic, thus rendering our system complete and coherent with respect to both analytical and synthetical knowledge. And this is the fundamental advance represented by this logic.

Since indeterminacy is absolutely central to physics and mathematics alike it is indeed fortunate that the true form of logic also embraces this category, particularly since, as we have seen, indeterminacy plays such an unexpected and elegant role in generating order. In philosophy too it allows us to outflank scepticism by lending scepticism a more precise, logical formulation. Indeterminacy also plays a key role in all the humanities and social sciences, albeit this role is less clearly defined.

As a matter of good practice we should always endeavour to solve problems of mathematics in a classical context before seeking to ascribe them to indeterminacy. Indeed a problem of mathematics cannot glibly be assumed to have a non-classical solution simply because we have not found the classical solution to it yet – as sometimes is suggested with particularly difficult and outstanding problems such as Goldbach's conjecture. In essence, for a mathematical problem to be accepted as possessing a non-classical solution we must objectively *prove* it to be indeterminate, much as Paul Cohen did with the continuum hypothesis. It is not sufficient simply to say; such and such a problem is difficult to prove therefore it is non-classical.

Nevertheless it is the case that, since many of the problems of mathematics have non-classical solutions then a deeper stratum to the axiomatic base (as defined by Zermelo, Fraenkel and others) must exist and its formal expression is indeed that of non-classical logic which must therefore be deemed to represent the foundation not just of "philosophy" but of the whole of the sciences including mathematics. Indeed, "philosophy" (conceived of as more than just an critical exercise) is nothing other than this set of relations.

This discovery of the correct logical foundations of all empirical and analytical knowledge had been the noble ambition of the logical positivists who failed, in my view, simply because they had inherited a limited and distorted idea of what logic actually is. Their foundationalist project, which was along the correct lines, was therefore sadly crippled by this erroneous conception.

It was only gradually, through the work of Brouwer, Heyting, Lukasiewicz and others that a more complete conception of what logic actually is was achieved. Notwithstanding the preliminary and somewhat isolated efforts of Dummett the natural extension to philosophy of this surprisingly quiet revolution in logic has not occurred.[121] Thus the major theme of logical philosophy since the war has remained the largely negative one developed by Quine and others of the failure of logical positivism. From this admitted failure philosophers (again led by Quine) have erroneously deduced the failure of foundationalism per-se and this sense of failure is the most significant hallmark of contemporary analytical philosophy.

Nevertheless, for the reasons given, this despair of foundationalism can be seen to be premature, particularly as its logical consequence is the abandonment of the idea of rationality per-se and hence of the sciences. The failure to accept these latter implications indicates, at best, a certain inconsistency and, at worst, outright hypocrisy. And yet, of modern philosophers perhaps only Paul Feyerabend has truly adopted this position, a position which, though consistent, I obviously believe to be unnecessarily pessimistic.

---

[121] Indeed, *Principia Logica* is perhaps best interpreted as the attempt to rectify this situation.

It is of course paradoxical that what we call "foundations" should be ungrounded or indeterminate. Yet this is the case precisely *because* logic is metaphysical and apriori and may ultimately be said to be a corollary of universal relativity.

In truth therefore, talk of "logical foundations" is merely a convenient (though entirely valid) metaphor since nothing is "substantial" in the Aristotelian sense. Logic is foundational merely in the sense that it is of greater *generality* relative to other, ultimately insubstantial elements of the overall system. It is only in this sense that we defend the concept of foundationalism.

77. The Illusion of Syntheticity.

The translatability of physics into geometry (therefore implying the dominance of the latter) also implies that physics (the dominant science) is reducible to the axioms of geometry and hence of mathematics. Since chemistry (and hence biology) are fully translatable into physics (with reference to Dirac's relativistic wave equation) it therefore follows that the axiomatic basis of *all* the empirical sciences must be identical to that of mathematics, even if the precise terms of such an axiomatisation never come to light. From this it follows that what we call the *synthetical* (as a fundamental category of philosophy) is ultimately a special case of the *analytical*. This is because physics (albeit laboriously) is reducible to geometry.

This, if true, obviously has important repercussions for our interpretation of the development of modern philosophy. For example Kant's central category of the *synthetic apriori* is made redundant except as a brilliant first approximation to this conclusion. This is because it becomes apparent (in light of the translatability argument) that the "synthetic" is "apriori" precisely *because* it is a disguised or highly complex form of the analytic. Thus Kant's view is affirmed but also significantly extended and clarified as well.

Also clarified is Quine's equally insightful critique of analyticity. It is because of the *complex* nature of syntheticity that the problems relating to analyticity arise. Nevertheless the concept "synthetic" is valid only when understood relativistically, i.e. as a complex or *special* case of the analytic, which latter category is therefore of more general or fundamental significance.

The universe is therefore a *geometrical* object (approximating to an hyper-sphere) and as such its characteristics must *in principle* be entirely reducible to their analytical elements. The fact that they are not is not indicative of the falsity of this view or of the reality of the cleavage between synthetic and analytic. Rather it is indicative of the *size* of the universe as a geometrical object. It is the practical problems animadverting from this fundamental fact which lead to the illusion of an absolute (and epistemologically devastating) cleavage between these two ultimately identical categories.

As we have seen the inherently "rational" nature of the universe (qua existent and qua geometrical object) is a simple product of what I have identified as the two engineering functions of the universe and their distinctive interaction. Being, as I have demonstrated, *apriori* in character, the engineering functions are also, ipso-facto, *analytical*.

78. Plato's Project.

What is perhaps most remarkable is that this superficially absurd view concerning translatability (of the sciences into geometry) is not entirely original, but represents perhaps the underlying, the true genius of Plato's philosophy. In the dialogue *Timaeus* for example physical reality (i.e. nature) is treated simply as an expression of geometrical form – both on the cosmological and the atomic scale.[122]

Thus although the specific details of Plato's vision are erroneous he nevertheless appears to anticipate the view (expressed in *Principia Logica)* of the relationship between physics and geometry perfectly. Not only this but, in privileging geometry as the dominant theory (relative to the sciences) Plato is able to subsume, in a deliberate fashion, two prior dominant theories – i.e. Pythagoreanism (which held that arithmetic was the dominant theory) and classical Atomism.

In Karl Popper's view it was the failure of Pythagoreanism, brought about by the discovery of irrational numbers such as $\sqrt{2}$, which inspired Plato to postulate geometry as an alternative dominant theory, particularly as irrational quantities represent no inherent difficulty in geometry. Thus the desire of the Pythagoreans to base not merely mathematics but cosmology itself in the arithmetic of the natural numbers was transformed, by Plato, into the desire to base them in geometry instead. This, rather than the erroneous "theory of forms" represents (according to Popper) Plato's greatest legacy to Western thought. Indeed, Popper shrewdly observes, the "theory of forms" cannot truly be understood outside of this historical and intellectual context.

Rather than the *theory of forms* the crowning achievement of the program initiated by Plato is, according to Popper, Euclid's *Elements*. This is because it is this greatest of works which represents the true model for all subsequent Rationalism and Formalism, as well as being the chief organon of the empirical method. Geometry indeed (as I have tried to argue) encapsulates everything (not excluding mathematics) *apart* from logic. As Popper cogently puts it;

> "Ever since Plato and Euclid, but not before, geometry (rather than arithmetic) appears as the fundamental instrument of all physical explanations and descriptions, in the theory of matter as well as in cosmology."[123]

What is perhaps most remarkable and most easily overlooked in Euclid's work is how empirically observable entities can be constructed out of entirely *apriori* geometrical principles. Thus, in effect, Euclid is able to construct an entire observable world out of nothing, much as modern mathematicians are able to generate the entirety of mathematics (including Euclidean *and* non-Euclidean geometry) and the continuum of numbers out of the empty set. In many ways it is *this* achievement which represents the terminus ad quem of Plato's remarkable and (almost) all encompassing project. And this is so not least because it indicates that the root of the synthetic (i.e. of nature) ultimately lies in the analytic (i.e. in geometry).

Plato therefore would have had no truck with the attempt (by Kantians) to place Rationalism and Empiricism on an equivalent basis. For Plato Empiricism would perhaps have been dismissed as a *method*, an instrument of Rational philosophy and nothing more. Through empiricism we are able to determine aspects of the (geometrical) form of our particular world, but through geometry, by contrast, we are able to conceptually construct, out of nothing, an *infinite plurality* of possible worlds. Rationalism is therefore clearly a more *general* basis for philosophy than Empiricism can ever hope to be.

---

[122] There is also the evidence of the famous inscription above the portal of Plato's academy; "Let no one enter who does not know geometry."
[123] Karl Popper, *The Nature of Philosophical Problems,* (IX). The remaining references in this and the next section are all to this remarkable lecture.

What is perhaps most important to observe about geometry as a theory however is that although it incorporates the sciences as a special case it does not ultimately point in the direction of either physicalism or materialism but instead points beyond these two peculiarly modern (and some would say barbaric) obsessions. For the essence of geometry lies neither in substance nor in arche but rather in relation and relativity. Thus Plato's project ultimately corroborates what I call universal relativity as against Aristotelian essentialism.

79. The Triumph of Rationalism.

Popper also interprets the sixteenth century renaissance in science as a renaissance in Plato's geometrical method, presumably (as according to my interpretation) in its *applied* form. He then goes on to cite quantum mechanics as a fatal counter-example to Plato's theory of translation.[124]

Quantum mechanics, according to Popper's view, is grounded in arithmetic rather than in geometry and he cites the arithmetical properties of the theory of quantum numbers as his support for this view. However, it is Popper himself who interprets number theory or Pythagoreanism as a special case of Plato's geometrical method. Consequently the arithmetical nature of quantum numbers is not itself in contradiction with the theory of translatability. What is more, Plato's view on this matter has since been confirmed in a spectacular manner by Cartesian coordinate geometry. Indeed, it was Descartes himself who demonstrated how the basic functions of arithmetic (addition, subtraction, multiplication and division) arise spontaneously as simple transformations in analytical geometry. Arithmetic is, in effect, geometry in disguise, as is the whole of number theory.

Furthermore, as I have pointed out elsewhere, quantum mechanics can properly be interpreted as a general theory grounded in the geometry of Hilbert space, just as Newtonianism is grounded in Euclidean geometry and Einstein's theory of relativity is grounded in non-Euclidean geometry. In each case the physical theory is discovered to be centered in a particular geometrical frame-work of yet greater generality.[125]

A final observation on this matter concerns the development of quantum mechanics since Popper's lecture. The dominant interpretation of quantum mechanics is now no longer the arithmetical one centered on the properties of the quantum numbers but rather a geometrical one based on invariance and symmetry transformations. This interpretation, known as gauge theory, illuminates the truly geometrical nature of quantum mechanics just as Boltzmann's statistical interpretation demonstrated the underlying geometrical significance of the laws of thermodynamics a generation before. Symmetry concepts are now as appropriate in talking about quantum mechanics as they are in discussing the General theory of Relativity where they constitute the axiomatic basis of that theory.

The implication of this (missed by scientists and philosophers alike) is that aposteriori physics is ultimately subsumed by *apriori* geometry as a special or applied (hence "aposteriori") case of the latter, precisely in line with the system described in *Principia Logica*. This view implies that the "synthetic" can only be "apriori" because (as Kant failed to realize) it is really the "analytic" in a complex form. In short, and counter-intuitively, nature is geometry in disguise.

---

[124] Ibid.
[125] It is also useful to recall that Boltzmann's statistical interpretation of the second law of thermodynamics effectively translates a theory about heat into a theory about order – i.e. a *geometrical* theory.

80. The Modern synthesis.

Although Rationalism implies the absorption of Empiricism (and of the sciences by geometry) as a special case of itself, Rationalism is itself subsumed by logic since it is logic which lies at the foundations of all mathematics, including geometry. Thus, as the progenitor of logicism Aristotle remains as fundamental a figure in the history of philosophy as does Plato.

The fact that Plato's academy soon came to be dominated by Pyrrhonic modes of thought is itself indicative of the fact that the character of logic (albeit not Aristotelian modal logic) is more fundamental than that of geometry. It is for this reason that we take the view that to be a philosopher is, first and foremost, to be a *logician*.

It was only after the Renaissance however that this informal hierarchy of philosophy came to be challenged by the advent of Empiricism whose character appeared, superficially, to diverge decisively from that of either Rationalism or Logicism.[126] And yet the founder of modern Empiricism (Francis Bacon) originally conceived of it as being rooted in a *complementary* species of logic to that underlying Aristotle's analytical philosophy. It was only with the advent of the so called British school of Empiricism (founded by Locke in the wake of Newton's achievements) that Bacon's fundamental insight (concerning the logical basis of inductivism) was lost and Empiricism came to be interpreted instead as a sort of *counter-example* to continental Rationalism and Logicism.

The intent of *Principia Logica* has therefore been to demonstrate that whilst this counter-example does indeed force a reformation in our understanding of Rationalism and Logicism (and indeed of Empiricism itself) it does not fundamentally alter the conceptual hierarchy first intuited by the ancient Greeks. In this case we find that the aposteriori sciences are *entirely* underpinned by apriori geometry which is in its turn *entirely* underpinned by apriori (non-classical) logic. More than this (which is everything) epistemology cannot say.

---

[126] It is true that, in some sense, empiricism along with logicism is a part of Aristotle's remarkable legacy to the world, but, as discussed earlier in this work, its distinctive formalization as a *rigorous* methodology and associated philosophy properly belongs to the modern period.

Appendix; Prime Numbers, Indeterminism and the Riemann Hypothesis.

The purpose of this paper is to prove the *inherent* indeterminacy of prime number distribution.[127] I would also like to suggest that a proof of the Riemann hypothesis follows as an *indirect* consequence of any valid proof of the indeterminacy of prime number distribution.

Orthodox attempts to prove the indeterminacy of prime number distribution have sought to do so by, in effect, proving that the non-trivial complex zeros of the Riemann Zeta-function all fall on the so called critical line. Consequently a proof of the Riemann hypothesis equates, ipso-facto, to a proof of the inherent indeterminacy of prime number distribution.

The *converse* of this would seem to be that (granted the prime number theorem) a proof of the inherent indeterminacy of prime number distribution equates to a proof of the Riemann hypothesis. If this is not a correct assumption then I can only apologize for my error. Its incorrectness would not however invalidate the very real discoveries described in this paper.

The key to proving the inherent indeterminacy of prime number distribution lies in the hitherto unnoticed link that exists between the prime number theorem and Boltzmann's equation for entropy. This connection lies in the fact that both formalisms make use of the natural logarithm $\log_e(x)$.

Indeed, in the case of Boltzmann's *statistical* interpretation the natural logarithm is nothing short of a measure of entropy – i.e. of disorder and randomness;

$$\frac{S}{k} = \log_e(x)$$

That is; as $\log_e(x)$ increases, disorder (S) also increases where (K) is a constant defined by the relationship between entropy (S) and the natural logarithm. This, despite its incidental usefulness to physics, is an apriori mathematical relation that effectively marks $\log_e(x)$ out as a perfect measure of disorder or unpredictability.[128]

Now it is a significant fact that $\pi(x)$, like the statistical measure for disorder, is *also* defined by $\log_e(x)$;

$$\pi(x) \approx \frac{x}{\log_e x} \approx Li(x) \qquad \text{(Which is the prime number theorem)}.$$

In effect therefore;

$$\log_e x \approx \frac{x}{\pi(x)} \approx \frac{S}{k}$$

---

[127] A proof of this is, in many ways, the latter-day equivalent of proving the irrationality of $\sqrt{2}$, at least in terms of the angry and defensive responses it draws from many so called professional mathematicians. All that is implied however is that the continuum of real numbers, though apriori in nature, contains, as a "special case", an aposteriori element, to wit the indeterminate distribution of the prime numbers.

[128] Entropy and Boltzmann's constant merely define particular *aspects* of $\log_e(x)$, they are not new or extraneous values.

And so;

$$S \approx k \frac{x}{\pi(x)}$$

In other words, $\pi(x)$ which is the number of primes up to and including a given number (x), is uniquely related to the *disorder* (S) associated with (x).[129] In effect, prime numbers represent the disorderly element in (x) such that (x), as we can see, *always* has a measure of entropy associated with it approximately equal to $\log_e(x)$.

To my mind this relation must necessarily be invariant across the continuum of positive integers (since it is a logical corollary of the prime number theorem). Furthermore it proves the inherent unpredictability of the individual prime numbers and hence, ipso-facto (granted the prime number theorem), the validity of the Riemann Hypothesis.[130]

This relation also accounts for the orderly growth of the function $\pi(x)$ which becomes, in effect, a representation of the geometrical expansion of entropy across the continuum of real numbers.
And this entropy only reaches a maximum at infinity; a fact which follows from what I have already shown coupled with Euclid's elegant proof that there exists an infinite number of prime numbers.
Incidentally, complex zeros are as unpredictably distributed on the critical line as are prime numbers on the continuum of real numbers. One could no more expect to find a non-trivial complex zero off the critical line than one could expect to find a prime number off the continuum of real numbers.
The alleged orderliness of the distribution of prime numbers is thus of a purely *statistical* nature, much like the "orderliness" in the distribution of random coin tosses. It is merely a consequence of what is sometimes referred to as the "law of large numbers". What is infact detected as orderliness is in reality nothing more than the *geometrical* nature of the expansion of entropy – i.e. of inherent indeterminacy.

The breakthrough in understanding therefore lies in recognizing that the different notations (lnx in the prime number theorem and $\log_e(w)$ in the Boltzmann equation) infact refer to the *exact same thing*, thus

---

[129] Note also the highly suggestive definition of any positive integer x provided by this equation;

$$x \approx \frac{S}{k} . \pi(x)$$

The entropy of (x) is clearly a function of $\pi(x)$, i.e. of the number of primes up to (x). And this is why the distribution of primes is inherently unpredictable since this unpredictability is what characterizes entropy. If prime distribution were predictable then no entropy would be associated with (x) and yet this is clearly not the case.
In truth, this result decisively linking the continuum to entropy (something previously only suspected and never accurately quantified) reinforces Gödel's conclusions in the light of the incompleteness theorem concerning the quasi empirical status of mathematical proof contained in; "The modern development of the foundations of mathematics in the light of philosophy", *Kurt Gödel, Collected Works*, Volume III (1961) publ. Oxford University Press, 1981.
This is because the distribution of individual primes cannot be known apriori but only aposteriori.
[130] Since the continuum of numbers is apriori this result indicates that entropy and indeterminism have an *apriori* character to them as well, something never previously suspected and certainly of groundbreaking epistemological significance. This major new theme (of the aprioricity of entropy) is expanded upon at greater length in my magnum opus *Principia Logica*.

allowing a fertile merger of these seemingly unconnected formalisms. Perhaps if Boltzmann had expressed his famous equation as; S=k.lnx , mathematicians might earlier have recognized its hidden relevance to the Riemann hypothesis.

Furthermore, the Riemann Hypothesis *itself* can be re-expressed as a precise statement about entropy, much as the prime number theorem can;

$$S = k \cdot \left(\frac{x}{Li_{(x)} + 0^{(x\frac{1}{2}\log x)}} \right) = k \cdot \frac{x}{\pi_{(x)}}$$

Which is itself an adaptation of Koch's well known expression of the Riemann Hypothesis.[131]

In conclusion, I have argued;

1) That the distribution of the primes must constitute the element of *entropy* (i.e. of disorder) that I have shown to exist in the continuum of real numbers. After all, to what could entropy in the continuum possibly refer to other than to the inherently indeterminate distribution of the prime numbers?
2) That the complex zeros of the Riemann zeta function constitute the element of entropy as it afflicts the continuum of *imaginary* numbers.
3) That positive entropy must continue to *increase* throughout the continuum – a fact which also follows (granted proposition 1)) from Euclid's proof of the existence of an infinite number of primes.
4) Granted propositions 1) and 3) and granted the prime number theorem it therefore follows that the Riemann hypothesis must be true.

A Proof of the P versus NP problem.

1) Due to the indeterminate character of prime number distribution it is therefore logical to suppose that the problem of factorization (concerning the identification of prime factors in polynomial time) is inherently insoluble.
 The indeterminacy of prime number distribution guarantees that the problem of factorization in polynomial time is inherently insoluble. This follows because a factorization method in polynomial time would ipso-facto allow us to identify prime number factors of arbitrarily large sizes in polynomial time, (the prime factors could easily be identified merely by repeatedly factorizing the factors identified by the hypothetical method.) Since this is prohibited by virtue of the inherent indeterminacy of the distribution

---

[131] Koch's version of the Riemann hypothesis is expressed (excusing script limitations) as the following equation;

$$\pi(x) = Li_{(x)} + 0^{(x\frac{1}{2}\log(x))}$$

From Koch's equation and bearing in mind our prior calculations it is a relatively simple matter to arrive at the restatement of the Riemann hypothesis as a statement about entropy just given. Yet this simplicity belies its great import.

of individual prime numbers it therefore follows that we have proven that the problem of factorization must be insoluble and that prime factors *cannot* be identified in polynomial time.

In other words; to solve the problem of factorization in polynomial time would ipso-facto entail the discovering of the whereabouts of *prime numbers* in polynomial time, which is inherently impossible. Therefore the problem of factorization is insoluble.[132]

2) In order for P to equal NP it would be necessary for the problem of factorization to be soluble in polynomial time and by non-random or *deterministic* methods. Since we have proven that this is not the case it therefore follows that $P \neq NP$.

3) The RSA code breaking problem (derived from public key cryptography) supplies an ideal counter-example to the hypothesis of P and NP equivalence. It will therefore serve to illustrate and confirm the general conclusion arrived at in proposition 2).

This follows because the RSA code breaking problem is known to be in NP.[133]

However, given the inherent impossibility of factorization in polynomial time as proven in proposition 1) it therefore follows that the code breaking problem *cannot* also be in P. Therefore $P \neq NP$.[134]

These results (concerning the Riemann hypothesis, the factorization problem and the P versus NP problem) combine to prove (incidentally) that the RSA public key cryptography system (presumably because it works with the grain of nature) is inherently secure. In short, no polynomial time procedure exists for breaking RSA type codes, they can therefore only be broken by chance or by brute-force methods.

We have also demonstrated that the P versus NP problem is a subsidiary of the factorization problem which lies at the heart of number theory and which we have solved (due to its intimate connection with the issue of prime numbers and indeterminacy) with a negative result. We may add that these negative results are in turn consequent upon our proof of the presence of entropy in the continuum of real numbers. Because of this presence the problem of factorization in polynomial time is not soluble and therefore P and NP cannot be equivalent.

Monday, May 30, 2005

Contact; A.S.N. Misra.
mnmhnmisra@aol.com

---

[132] At the heart of the P versus NP problem is the problem of factorization and at the heart of the problem of factorization is the problem of the indeterminate distribution of prime numbers. It is *because* of this latter difficulty that we can say that P does not equal NP.

[133] This is common knowledge and so a rehearsal of the proof that RSA is in NP is not required here. All that is required is a proof that RSA *is not* in P.

[134] Another way of expressing this is to say that for the RSA code-breaking problem to be solvable in P would entail a solution to the problem of factorization in polynomial time, which would in *its* turn entail the *deterministic* distribution of the primes. We have already proven that the prime numbers are *not* deterministically distributed. Therefore this problem (which *is* in NP) cannot also be in P.